T0211613

Women, the Novel, and Natural Philosophy, 1660–1727

Women, the Novel, and Natural Philosophy, 1660–1727

Karen Bloom Gevirtz

palgrave
macmillan

Softcover reprint of the hardcover 1st edition 2014 978-1-137-38920-6

First published in 2014 by
PALGRAVE MACMILLAN®
in the United States—a division of St. Martin's Press LLC,
175 Fifth Avenue, New York, NY 10010.

Where this book is distributed in the UK, Europe and the rest of the world,
this is by Palgrave Macmillan, a division of Macmillan Publishers Limited,
registered in England, company number 785998, of Houndmills,
Basingstoke, Hampshire RG21 6XS.

Palgrave Macmillan is the global academic imprint of the above companies
and has companies and representatives throughout the world.

Palgrave® and Macmillan® are registered trademarks in the United States,
the United Kingdom, Europe and other countries.

ISBN 978-1-349-48230-6   ISBN 978-1-137-38676-2 (eBook)
DOI 10.1057/9781137386762

Library of Congress Cataloging-in-Publication Data is available from the
Library of Congress.

A catalogue record of the book is available from the British Library.

Design by Newgen Knowledge Works (P) Ltd., Chennai, India.

First edition: March 2014

10 9 8 7 6 5 4 3 2 1

For Naomi, Ben, and Stephen

Contents

Acknowledgments   ix

Introduction   1

One   Notions of the Self   15

Two   An Ingenious Romance: The Stable Self   35

Three   The Fly's Eye: The Composite Self   71

Four   The Detached Observer   101

Five   The Moral Observer   127

Conclusion   167

Notes   173

Bibliography   217

Index   239

Acknowledgments

Thanking the many people who have made this book possible is one of the great pleasures of such a project. It gives me considerable sympathy for my authors, who in their different ways rejected the notion of isolated genius and recognized that the interaction between individual and context is necessary to creation.

The cover image, "Self-portrait" ("Selbstbildnis") by Angelika Kauffmann (1780–81), is used with permission of the Bundner Kunstmuseum Chur. I wish to thank Nicole Seeberger for guiding me through the process with the museum. Parts of chapter 2 appeared in "Aphra Behn and the Scientific Self," in *The New Science and Women's Literary Discourse: Prefiguring Frankenstein*, edited by Judy Hayden (2011, Palgrave Macmillan) and are reproduced with permission of Palgrave Macmillan.

Seton Hall University provided resources that enabled the completion of this study. Dr. Larry Robinson, Provost of Seton Hall University, granted a sabbatical. The University Research Council awarded a summer stipend and then-Interim Dean Joan Guetti secured additional funding for travel. Mary Balkun, Chair of the English Department and the members of the English Department provided funding and approved time for my research. Rebecca Warren, Department Secretary, offered good cheer and administrative support.

A number of colleagues at Seton Hall have also been integral to my work. Dermot Quinn and James McCartin (now at Fordham University) inspired me with their erudition and love of learning; they made Bacon, Hobbes, and Locke wondrous and thrilling. Anthony Lee, Marta Deyrup, Lisa Rose-Wiles, Martha Loesch, and Mabel Wong at Walsh Library exemplify generosity, resourcefulness, and determination. Jonathan Farina, Donovan Sherman, and Angela Weisl heroically read chapters and offered keen commentary and suggestions. Last but far from least, thanks to Anthony Sciglitano, Williamjames Hoffer, Amy Hunter, and James P. Kimble.

I am grateful to many colleagues beyond Seton Hall. Rupert Baker, Library Manager at The Royal Society, provided thoughtful, invaluable assistance. The librarians at the University of Chicago Library and the New York Public Library, including the Rare Books and Manuscripts Division, were helpful and kind. Maureen Duffy, Rachel

Carnell, Jessica Munns, Al Coppola, Judy Hayden, Margarete Rubik, Martha Bowden, Rivka Nelson, and Patrick Spedding (by a roundabout route) shared their expertise and offered me invaluable opportunities to refine my thinking. Judith Zinsser gave priceless advice and mentoring. Jane Spencer made crucial observations and suggestions and generously volunteered a copy of her forthcoming work on Mary Davys. Elaine Hobby, Claire Bowditch, and C. J. Flynn-Ryan of Loughborough University and the Aphra Behn Society of Europe created a conference at which much of the methodology of this study was reimagined and affirmed. I cannot express the extent of my gratitude to Mona Narain, Aleksondra Hultquist, Cynthia Richards, and Mary Ann O'Donnell. Everyone, especially they, have tried diligently to preserve me from crimes of argument and prose, but no doubt I have still managed to commit at least a few.

Many thanks to the friends and family who shared the journey with me. Among them, Emery Weinstein was an excellent research assistant. Robert Malenka told me that scientists would like my argument; Ruth and Bert Malenka have encouraged me throughout. The Blooms, Weinsteins, and Gevirtzes have been unstinting with their love and support, particularly my parents, Susan and Steven Bloom, and in-laws, Ed and Marilyn Gevirtz.

Naomi and Ben have patiently awaited the end of this book project and the last promise that I would be there in just one minute. Stephen, my *bashert*, makes all things from the mundane to the celestial possible. For those reasons and an infinitude of others, they are inspiring and miraculous. To them and for them, thanks are due.

Introduction

It was a great while before she could recover from the Indisposition to which this fatal and unexpected Accident had reduced her: But, as I have said, she was not of a Nature to dy for Love...
 —Aphra Behn, *Love-Letters Between a Nobleman and His Sister*

But her sanguine Temper soon dispell'd the Mist that would have clouded her warm Imagination, and she was resolved to hope Sir John would like a College-life so well, that some Years would drop before he came again.
 —Mary Davys, *The Accomplish'd Rake*

In 1687, Aphra Behn published the third installment of *Love-Letters Between a Nobleman and His Sister*, in which a narrator fully knowledgeable about the characters' most intimate inner thoughts and most private doings tells their story infused with her own physical, political, and social desires. For Behn, it was inconceivable that anyone could distinguish what today we might call one's subject position from one's knowledge. Forty years later in *The Accomplish'd Rake* (1727), Mary Davys created an embodied narrator also fully knowledgeable about the characters' most intimate inner thoughts and most private doings, but this narrator's material existence was essentially irrelevant beyond its fact. Davys's narrator's knowledge was validated by a detachment that Behn would have treated as a delusion.

The difference between these narrators testifies to more than individual authorial preference and imagination. The decades between the start of Behn's career in narrative prose fiction and the end of Mary Davys's novelistic career were characterized by extensive, vehement debates about the nature of the human capacity for knowledge and the appropriate language for representing that knowing self. These debates were instigated by the revolution in natural philosophy—formerly called the "scientific revolution"—and quickly carried into popular culture, often with the aid of natural philosophers such as Robert Hooke, Robert Boyle, and John Locke. There was no rhetoric for these ideas of human capacity or for representing and performing new ideas of the act of knowing, including literary conventions or genres. Even if there were extant conventions or forms that could support or execute the values of the philosophical revolution, the combination of new pressures, new ideas, and new rhetorical needs rendered those existing forms problematic or even obsolete. *Women, the Novel, and Natural Philosophy, 1660–1727* shows how professional writers, particularly early novelists, seized on these debates about the nature of the self and the language used for depicting that self. These writers used, tested, explored, accepted, and rejected ideas about the self in their works to represent the act of knowing and what it means to be a knowing self. As they did so, they developed structures for representing authoritative knowing that contributed to the development of the novel as a genre and to literary omniscience as a point of view.

By "self," I mean the combination of body and mind that interacts with knowledge, whether that knowledge comes from authority or experience. This definition is somewhat broad because the definition during the late seventeenth and early eighteenth centuries was somewhat broad, in no small part because ideas about the self were developing during the period. This study therefore is not about subjectivity nor is it about the individualistic or romantic self. Those entities require both a dominance of Lockean ideas and a dismissal of the flesh that did not exist during this period. Charles Taylor, Timothy Reiss, and Dror Wahrman, for example, have shown that the idea of the "self" changed considerably over the course of the early modern period and not in a tidy linear evolution from one notion to another.[1] Amos Funkenstein's *Theology and the Scientific Imagination* established the connection between systems of knowledge and systems of identity construction, describing how through the seventeenth century, the "scientific imagination" or the epistemology of science affected ideas, including those rooted in theology, of what humans can know and how they can know it.[2] Although thinkers recognized that contextual factors influenced the self by interacting with it, over the seventeenth and into the eighteenth century the self was increasingly understood to have its own, however influentiable, identity separable from context.[3] Similarly theologians, philosophers, and the like argued whether the old definition of the self as composed of body, mind, and soul still held and if so, what each components' function was, a debate that raged on into the 1730s. Eventually humans were understood to possess a "well-defined, stable, unique, centered self," as Wahrman puts it, but that came later in the eighteenth century.[4] It is the fact of this debate during this early period as well as aspects of it that concern my work here.

In crucial ways, the heart of this discussion beat in natural philosophy. The anachronistic term "scientific revolution" has been replaced by the more historically accurate phrase "philosophical revolution" as historians of science such as Simon Shaffer, Steven Shapin, Michael Hunter, Peter Dear, Judith Zinsser, Lynette Hunter, and Sarah Hutton have documented and described the period.[5] Their work has complicated the triumphalist narrative of the rise of the Royal Society and with it, the rise of empiricism.

My study therefore eschews the term "scientific revolution" for "philosophical revolution," although occasionally the phrase "the new science" will appear, meaning "science" in the older sense to evoke the idea of a new way of approaching knowledge and the act of knowing. The revolution in natural philosophy was a catalyst for the creation of some modes of thought as well as a force against which other modes reacted. Natural philosophy, it should be remembered, merged seamlessly with political philosophy, which in turn overlapped with other developing branches of thought such as economic philosophy and theology. As such, it had ideological as well as epistemological and literary influence. Admittedly, the opening salvo of the philosophical revolution is hard to pin down. What seems fairly safe to say is that by the time Francis Bacon was writing his *Great Instauration* (1620) and Galileo was publishing *The Assayer* (1623) and *Dialogue Concerning the Chief World Systems* (1630), significant changes in thinking were taking place. These included a profound suspicion of the body's perceptions and the mind's productions, an embracing of technology such as microscopes and air pumps as a way of remedying these flaws of the self, and an interest in reconstructing epistemology and communities in which knowledge was produced and validated.

In England, the philosophical revolution was well underway by the late seventeenth century. Groups of thinkers such as the Cambridge Platonists, the Greshamites, and the Invisible College were establishing themselves in the key educational centers of Oxford, Cambridge, and London. They were talking with each other, forging new connections in England and on the Continent, and exploring a variety of methods and ideas about the self and how it should be managed in the production of knowledge. Thinkers argued a variety of positions on whether and how much of the cosmos could be known by humans, and what role there was for God in emerging notions not just of nature but also and perhaps especially of the human self. By the 1680s, ferocious debates such as those about the use of technology, the criteria for acceptable witnesses, or the existence of miracles and spirits had generated a multiplicity of views comprising a context in which English society and culture would develop over the next several decades.[6] Sixty years later those debates were beginning to settle into certain grooves and produce certain results, particularly in the epistemology underpinning English society and its changing institutions, but it is important to remember that those debates continued. Grounding arguments about knowledge and method was a new notion of the self, a "breathtaking sense of human potential, a belief that everything in the world might be made more rational and logical," as Michael Hunter puts it.[7] The key here is that epistemology, which encompasses the method for generating knowledge and the definition of knowledge, reflects or responds to notions of the self: ideas about human capacity produced an epistemology during this period, and with it, the need for new uses of language, new modes of writing, and new genres.[8]

The definition of the self and the delineation of its capacities were central to this "breathtaking sense of human potential." For some thinkers, such as Francis Bacon, Robert Boyle, and Robert Hooke, body and mind were essential components of the self. Whatever the flaws of the body, they were necessary to—or at least unavoidable parts of—the self. For thinkers like John Locke, the self could not exist without both body and mind, but the former was far less important than the latter. For other thinkers such as Descartes and his followers, the self in the form of the rational mind was lodged in the body but could decipher the world without the body's help. Critics of the "new science" recognized how crucial and how problematic these ideas about the self and the new science's dependence on them were. This idea of the experimenter was a significant sticking point for Margaret Cavendish, Duchess of Newcastle, for example. As Anna Battigelli explains it, Cavendish's "concern lies with [experimental philosophers'] unwillingness to acknowledge the inevitable interference of their own subjectivity."[9] Arguments about the nature of self, whether and what knowledge was possible for that self, and how it could all be expressed through writing continued well into the eighteenth century. This study shows how this contentious, varied discourse about the self offered a rich environment of ideas and rhetoric for early novelists, who used their narratives to engage in the debate and who pioneered not just ideas about the self but also techniques for writing about and representing it.

These debates about the nature of the self and the language used for depicting that self and its functions were taken up by early novelists, people who were not philosophers but who were very interested in the philosophical questions swirling around them. These writers adopted, interrogated, and rejected ideas about the self in their works as the subject or in the characterization of their players, and also in the use of narrative to represent the act of knowing and what it means to be a knowing self. Narrative structure takes shape through interactions between text and ideology: the "authority of a given voice or text is produced from a conjunction of social and rhetorical properties," Susan Sniader Lanser explains.[10] Historically speaking, it was not necessary to have read works of natural philosophy, not even Newton or Locke, to be familiar with the issues underpinning the English philosophical revolution in the late seventeenth and early eighteenth centuries. George S. Rousseau points out that natural philosophy entered the "literary imagination" even of authors who did not read the source texts, such as some of the "Newtonian" poets whom Nicholson describes.[11] Ideas about the nature of the self were everywhere and they provided excellent grist for the literary mill, including the novel. According to Michael Mascuch, constructions of the autobiographical subject changed considerably during this period.[12] More recently Rachel Carnell and Jesse Molesworth, following Ros Ballaster, have shown ways in which the emerging novel engaged with epistemological issues through literary and rhetorical form and consequently shaped and was shaped by epistemological debate.[13] My work extends their insights, infusing this approach with a recognition of gender's role in the interaction of structure and epistemology.

It was particularly significant for women that ideas about the self passed readily between natural philosophy and the cultural mainstream. Women's access to education and beyond it, to certain kinds of information, skills, and ideas, was troubled during the late seventeenth and early eighteenth centuries. At first glance, the philosophical revolution would seem to have offered women unprecedented access to knowledge and the means of its production. Londa Schiebinger and Erica Harth, for example, show how women participated in natural philosophy during the seventeenth century.[14] Until the mid-eighteenth century, women were encouraged to interact with natural philosophy by becoming proficient in math, read the rapidly growing number of popularizations of philosophical works, and attend open lectures.[15] Women were also encouraged to participate in natural history well into the eighteenth century.[16] Aristocratic and royal women in England during the seventeenth and early eighteenth century had even better access to the productions of natural philosophy.[17] Access is very different from approval, however. Women were never welcome in natural philosophy's formal or institutional settings and they became more unwelcome as the revolution progressed.[18] Despite popularizations, women were less and less able to secure the kind of education that would enable them to understand, let alone participate in philosophical discourse and development. Judy Hayden observes that "While women's demands for a 'masculine education' were no doubt alarming, they were also largely ignored."[19] By the end of the eighteenth century, women like Emilie du Châtelet, however impressive, were considered unusual or "freakish," worthy of admiration perhaps but definitely not to be emulated.[20]

Furthermore, women had less opportunity to "do" natural philosophy. As Elizabeth Potter puts it, "modern experimental science was intensely gendered at its inception."[21] Women faced social opprobrium, practical issues such as lack of money, time and space, and the deficiencies of female education, not to mention the absence of a model for a female philosopher in the cultural or philosophical imagination. The individual philosopher and his colleagues were resolutely male. Knowledge was generated by male activities, in male spaces, by male selves. A woman could get a "Newton for the Ladies" but she could not have a laboratory; she could have a greenhouse but she could not have an observatory; she could collect shells at Weymouth but she could not collect them with Captain Cook in Tahiti.[22] The universities and the Royal Society were closed to women. In England, a woman could consume what others considered knowledge but if she produced or tried to produce her own she faced significant obstacles and censure. Doubtless, exceptional women did it but not remotely in the same numbers, to the same effect, or under the same circumstances as men. And that exclusion from the modes and means of knowledge production, including the society of other—male—philosophers, was and remains a key component to women's ability to make knowledge and to shape epistemology.[23] A woman was not admitted to the Royal Society until 1945; currently, women make up only six percent of its Fellows.[24] During the late seventeenth and early eighteenth centuries, then, it was consequently of considerable importance to women that ideas from natural philosophy entered the cultural mainstream, whether identified as originating in natural philosophy such as a simplified version of Newton's writing aimed at female readers, a cheap pamphlet on the longitude crisis, or ideas in social circulation. Generally speaking, a woman could not publish a book of experimental philosophy, but she might write a novel engaging with its ideas.

At the same time, knowledge or methods of knowledge acquisition associated with women were devalued. Women were "depleted of epistemological agency," in Donna Haraway's words, and the new ways of knowing were rendered superior and male.[25] In its early years, as Frances Harris and Lynette Hunter have shown, male experimental philosophers justified their work by distinguishing it from women's domestic empiricism to the latter's detriment.[26] In elevating "propositional knowledge, or 'knowing that'" over "'Knowing how,' or skilled activity," natural philosophers "creat[ed] a hierarchy of knowledges" that left women subordinate or inferior and therefore worthy of exclusion.[27] Pushing women out of empirical practices and claiming what had traditionally been associated with the female domestic for the male intellect reveals that men like Boyle valued male and female existence and experience differently, and that the rational, unified, objective self they were constructing had to be a product of the male realm and resolutely male. With new modes of knowing unfriendly to women and their experience, and old modes of knowing devalued or invalidated, it is hardly surprising that women would be interested in the epistemological changes, particularly the changes affecting what they could be said to know and how. How new and old ideas interacted, whether in the more exclusive realms of natural philosophy or the more accessible realms of popular culture, would have been compelling for women attracted to any aspect of the discussion.

Debates about the self were therefore significantly inflected by gender, whether explicitly or implicitly. Women and men were interested in such questions, sometimes in the same ways and to the same extent. But for the women authors who grappled with the emerging concept of the knowledgeable self, these questions had other dimensions worth considering as they affected the construction of modern epistemology. Gender therefore has a highlighting function in this study, rendering more acute and more visible issues involving the self in the developing epistemology of the time. The role of gender in constructing the self meant that gender also had a role in constructing that self textually. Women authors did not always or inevitably reject ideas in circulation; those ideas offered an infinite number of positions and concepts to adopt and test. Women novelists were not the only writers interested in developing ideas about the self, but because women had more to lose in the development of these notions of self, the women writers who engaged with these issues rendered highly visible the challenges of managing the body's contribution to the self. When handled by women writers interested in such issues, gender served as a coloring agent as it were, helping to reveal aspects of the debates of the time and thereby helping to reveal the role of such issues in the development of novelistic structures.

Issues of access and opportunity made the novel a particularly useful site for such investigations. New ideas about the self put pressure on existing genres, whose conventions for representing the

self were appropriate for older rather than newer ideas about the
nature and capacity of the self. The resulting struggle to find modes
of writing appropriate to ideas about the self characterizes much
philosophical writing from the period, as historians of science have
demonstrated.[28] "[T]he changes in literary form and the changes in
the effective meanings of authority and experience are firmly integrated," Peter Dear writes, adding, "They may not be separable at
all, but together may be seen as constituting a new structure of natural philosophical discourse and practice."[29] Susan Sniader Lanser has
made a similar point about narrative fiction, arguing that literary
techniques and especially point of view are constructed in relation to
notions of authority constituted within an ideological framework.[30]
There was no demarcation between "natural philosophical writing" and narrative during this period; Robert Boyle, for example,
employed narrative and personal techniques in order to link his
method and conclusions to the body and character of the experimenter, and to position his readers as "virtual witnesses."[31] Literary
forms that allowed or even facilitated exploration, investigation, or
inconclusiveness offered an advantage to an author with these ideas
in mind. Without a set of conventions and rules but with a developing readership, the early novel was a fine site for attempting methods
of representing and conceptualizing the self. The novel also offered
opportunities to female as well as male authors; it was particularly
available to women in ways that other genres were not.[32] This acces-
sibility to women, whatever the genre's disrepute, made it a particularly active site for the development of ideas about the self and of
literary forms for representing and expressing those ideas. Although
scholars have looked into the relationship between literature and
natural philosophy, fewer studies, such as those by J. Paul Hunter,
Michael McKeon, and Rebecca Tierney-Hynes, have addressed the
relationship between natural philosophy and the novel as a genre.[33]
Women, the Novel, and Natural Philosophy expands that area of
inquiry and also breaks new ground by examining the confluence of
genre in structural terms, natural philosophy, and a third key factor,
gender.

   The use of the term "novel" for the works discussed in the following chapters may demand some defense. Battles are still being waged
over the definition of the term "novel" and over the origin of the
genre. The Heirs of Watt look to Defoe, Richardson, and Fielding
for the sources of the genre and its definition. Their challengers
might claim Aphra Behn, Delarivier Manley, or Eliza Haywood as
the first novelists; might reject the concept of an eighteenth-century

novel altogether by arguing that everything before Scott and Austen's work was a prenovel or protonovel; or might see Behn and Congreve, Manley and Defoe all together producing the novel. I grant that the novel of Jane Austen and Walter Scott is not the novel of Eliza Haywood and Mary Davys. I also agree with J. A. Downie that to speak of the novel before 1740 is to speak of a collection of narrative prose fictional forms with overlapping sets of characteristics rather than one monolithic category.[34] My point in this book is to show how authors of narrative prose fictions before 1730 drew on or created a plethora of strategies in response to certain key questions about what humans can know and what the quality of human knowing is. To distinguish the kinds of innovative, creative, and also different narratives from the arguably more formally coherent narratives of the late eighteenth century, and to focus on this quality in those pre-1730 narratives, I am calling them "early novels." Why stop at 1727? Coincidentally, it is the year Newton died and Mary Davys published her last original novel.[35] More substantially, the 1720s mark the end of the first period of ferment and creativity before a lull in novel production in the 1730s, which was followed by the emergence of what scholars generally agree is definitely the novel in the 1740s. That is not to claim something special about the 1720s for the development of the novel, as John Bender and Paula Backscheider have suggested, but rather to point out with Downie that this decade is the last in a series of decades in which innovation of a certain kind was taking place; 1727 marks the end of an arc, as it were.[36]

In short, early novelists were grappling with ideas about the self available in the cultural mainstream but originating in the philosophical and epistemological revolutions of the seventeenth and early eighteenth century. Women authors' engagement renders particularly visible the problems inhering to these ideas of self and knowledge and the challenges to rhetoric that those ideas posed. Carnell has argued that "One reason for the continued marginality of these women writers in the eighteenth-century canon...is that their work has been analyzed most frequently in terms of political or cultural history, rather than in terms of the development of the novel's formal structure."[37] Extending her demonstration that the conventions of narrative realism were forged through the conflict between what Carnell calls "different partisan versions of human experience and of the normative political individual," I show that novelistic conventions for representing authoritative, detached knowing were also forged through those same conflicts at the level of epistemology, not just the political debate inflected by epistemological conflict.[38] Their structural engagement and innovations produced representations of and performances of the knowing self and the act of knowing.

These innovations also contributed to the development of literary omniscience, what is often called "the omniscient point of view." Omniscience has been considerably theorized but much less often understood in historical terms. Narrative theorists, among them Gérard Genette, Jonathan Culler, and Nicholas Royle, have long argued over the definition, even the existence of literary omniscience precisely because the term encompasses so diverse an assortment of attributes.[39] As Paul Dawson explains, however, the problem is not categorization but historical insensitivity: "While narrative theory acknowledges historical shifts in fashion, it operates with a synchronic understanding of omniscient narration as a static element of narrative, produced by the structural relationship between focalization and voice." Narrative theory depends on an understanding of "omniscience" based on the nineteenth-century novel, which is not only different from the eighteenth-century novel but also different from the early stages of the eighteenth-century novel.[40] *Women, the Novel, and Natural Philosophy* takes up the two challenges that Dawson poses to literary analysis: to understand literary omniscience as a rhetorical performance that readers accept as a representation of omniscience, and to acknowledge the impact of cultural or historical context on the conventions that comprise that performance.[41] The recognition by critics such as Michel Foucault, Judith Butler, and Ellen Spolsky that language constructs and controls aspects of identity applies to constructions of point of view, as Lanser has argued.[42] Readers accept a set of rhetorical conventions as a performance of omniscience, not as the thing itself, and they have to be trained to accept such conventions as a representation of omniscience. This study proposes that this training began during the late seventeenth and early eighteenth centuries in works invested in the epistemological debates.

Viewed this way, inconsistency in the category "omniscience" is wonderfully revealing of the emerging novel. As Dawson notes, "The formal contingency of omniscient narration results from the fact that its narrative authority relies upon historically shifting literary-cultural conditions which determine the status and function of the novel in the public sphere."[43] Audiences accept certain collections of rhetorical conventions as a performance of "omniscience." Those collections are historically dependent: Whatever that culture, at that moment, chooses to accord the authority of full, reliable knowing is what comprises omniscience. Hence, a study of the relationship between the literary conventions that comprise full, reliable knowing and the epistemology of a given moment illuminates not only the epistemology and the phenomenon of literary omniscience but also the nature of the novel as a genre at that time. How the eighteenth century constructed omniscience "in terms of the specific claims for cultural authority which enable this narrative voice to function," as Dawson puts it, sheds light on ideas of knowledge, of human capacity to know, and on the role of the novel in demonstrating and representing those ideas and that capacity.[44]

What then can historical context teach us about epistemology, literary omniscience, and the emergence of the novel in the late seventeenth and early eighteenth centuries? John Bender has offered one answer, arguing in *Imagining the Penitentiary* that the novel's preoccupation with observation and punishment produced the omniscient point of view in literature and with it, Western culture's concepts of what it means to know. Cognitive literary theorists have proposed another, imagining that British thinking entered overnight into Lockean empiricism and consequently assumed a focus on mental, particularly cognitive, states in narrative.[45] Neither "rise" narratives nor single-entity triumphalist narratives adequately describe epistemological or generic developments during this period, however. As I demonstrate, some of the techniques that were eventually gathered under the heading "omniscience" appear there because they are performances, ways of representing authoritative, reliable knowing developed in response to the ideas and challenges coming out of the revolution in natural philosophy. The chapters of this book show that the techniques we have put under the heading "omniscience" were ways of recapturing and representing, of performing, a self in action. There is no one set of conventions for omniscience precisely because a constellation of techniques was born from the same urgent need to create the sense and performance of knowing.

Chapter 1 describes in greater detail the notions of the self supporting the revolution in natural philosophy during this period. The philosophical revolution depended on certain ideas about the self, including stability, unity, and capacity to compensate for inevitable flaws of perception and reason. These ideas were discussed, debated, and challenged by a variety of writers in a variety of ways; they did not appear fully formed from the mind of John Locke, and they were not accepted readily by philosophers or the hoi palloi. In fact, as chapter 1 reveals, Locke's ideas about the self were part of a larger discussion. While his contribution to that discussion is far-reaching, to put it mildly, he was hardly alone in considering these issues and grappling with the questions about what if anything limited the self in its quest for full, reliable knowledge about nature. Chapter 1 thus sets the historical, literary, and epistemological stages for the analyses of literature that follow, establishing the crucial fact of the debate and discussion about the self as well as the terms on which these debates and discussions depended.

Chapters 2–5 show how debates about the nature and capabilities of the self shaped the creation of novelistic structures for representing knowledge and knowing during the early decades of the novel's development in England. Chapters 2 and 3 examine the idea of the fallible self and its implications for acquiring reliable knowledge. Chapter 2 focuses on Aphra Behn's narratives. *Love-Letters Between a Nobleman and His Sister* (1684–87) was published before Locke's work but during—and in response to—a period in which natural philosophers wrote extensively about the self and its capacities. *Love-Letters'* formal evolution through its the three sections reveals how Behn developed narrative structures to interrogate the human capacity for knowledge, reflecting her increasing concerns with the social and political implications of ideas of selfhood gaining ground during the 1670s and 1680s. This chapter concludes with an examination of *Oroonoko*'s (1689) challenge to ideas of stability and self-knowledge. Behn's work exemplifies my argument that the early novel emerged through its engagement with these questions, and my accompanying point that the narrative techniques for representing point of view did not result solely from an interaction with specifically Lockean ideas but with broader debates about the self, under whose umbrella Locke's ideas fall.

The next three chapters consider works more clearly identifiable as early novels, looking at ways in which aspects of the debate about the self provoked structural innovations to narrative that helped establish the novel as a genre. Published in the decades following the publication of Newton's *Principia* and *Opticks* and the second edition of Locke's *Essay Concerning Human Understanding*, these early novels show how ongoing debates about the nature of the self, some in response to Newton and Locke's notions, created a rich, supportive environment for structural innovations in narrative. Chapter 3 focuses on Jane Barker's Galesia trilogy (1713/1719, 1723, 1726). Barker was trained in the "new science" by her brother; her works reveal a familiarity with ideas, issues, and knowledge produced by the philosophical revolution both in their content and, as I argue in this chapter, in their structure. Barker's increasingly complex framing techniques reflect a belief in a limited, flawed self and propound a communal approach to knowledge-gathering as a solution to the problems of individual limitations. Barker's ever-receding narratorial perspective approaches the entirely removed omniscient point of view in later novels, but her point that a single perspective cannot achieve full knowledge means that her framed narratives always point to their own insufficiency.

Chapters 4 and 5 look at narrative structures responding to the increasing tension between individualism or detachment and socialization or integration. While Behn and Barker place the insufficient self at the center of their narrative structures, the next two writers in this study, Eliza Haywood and Mary Davys, place the relationship of the self to its context in the generation of knowledge at the center of theirs. The titles of these chapters thus shift from the term "self" to "observer," which implies a connection between self and context. Chapters 4 and 5 are concerned with the question of morality and knowledge, a profound problem for natural philosophy during this period and particularly for women, who were increasingly constructed as moral forces. In the early novels discussed in these chapters, narrators attempt to reconcile the seeming contradiction in natural philosophy's view that the self needs to be present but otherwise disengaged for knowledge to take place, both in the self's acquisition of it and in the sharing of it. In different ways, Haywood and Davys pioneer representations of detached, full knowing that provide techniques for the omniscient point of view. Their narratives, although contemporaneous with Barker's, show the influence of debates about Lockean notions of the self. They also engage with the problems of moral value arising out of the premium placed on detachment, disengagement, and independence from previously received authority.

Chapter 4 examines Eliza Haywood's *The Tea-Table* (1725) and its exploration of the role of detachment and integration in the generation of valid and morally valid knowledge. Rarely discussed and then usually in the context of Haywood's break with Aaron Hill and his circle, *The Tea-Table* proves a thoughtful consideration of the relative values of independent critical thinking and social integration and awareness in the generation and dissemination of knowledge. Haywood's narrative considers the value of detachment primarily as a social issue: the narrator moves between being separate from his society and being part of his society, allowing Haywood to investigate the role that isolation plays in the generation of knowledge, particularly knowledge with a high use-value like moral knowledge.

Similarly, chapter 5 shows how Mary Davys' novels also investigate the relationship between being integrated into a group and being independent of a group in the production, validation, and distribution of knowledge. This chapter offers the first analysis of all of Davys's novels, from *Alcippus and Lucippe* (1704) to *The Accomplish'd Rake* (1727), as a body of work. It demonstrates that over the course of nearly three decades, Davys used and developed forms of narration attempting to reconcile the knowledge offered by being an outsider with the valuable and valid moral judgment offered by being socially integrated. Davys's novels reveal that she also is interested in the idea, most famously propounded by Locke, that the mind or intellect can be detached from the body. Her most complex narrators are separate from the events and characters that she describes, but their sensory perceptions are also rendered unimportant to the narrator's knowledge beyond the fact that the body that generates perceptions exists. Like the narrator of Haywood's *Tea-Table*, Davys's narrators contain conventions for representing knowing that would come to be associated with the omniscient point of view.

The revolution that gave the Western world gravity, the circulation of blood, and constitutional government was also responsible to some extent for the novel and the rhetorical forms that readers came to accept as the representation of our selves. New definitions of knowledge and the quest for methods to obtain it required new conceptions of human capacity. Contested, accepted, and rejected, those conceptions of the self and the modes of writing that they necessitated inspired new forms and structures, contributing not only to the creation of the novel as a genre but also to techniques for representing reliable, full knowing, techniques whose descendants are in use today. For women, increasingly excluded from notions of a self imbued with reason and agency and from opportunities to modify the epistemology depending on those notions, much was at stake. Narrative's concurrent investigations of the human capacity to know and of text's role in writing that self and that action thus reveal the interactions of literature and science, text and epistemology, that helped shape the novel and our modes for representing the self in narrative.

Chapter One

Notions of the Self

The Scientific or Philosophical Revolution depended significantly on a new notion of the self. Crises of "knowability," that is, challenges to established definitions of knowledge and established methods for identifying and transmitting it during the early modern period were driven in part by new ideas of what human beings can achieve.[1] Philosophers developed new schools of thought predicated on the idea that the limits of human knowledge were the fault of human error: errors of the body, of the mind, and of the methods for gathering knowledge. Capable of acquiring an accurate understanding of the workings of nature, the self simply needed the right combination of training, tools, and technique. In this view, humans are not limited to certain truths by our place in the cosmos. Certainly philosophers of the revolution such as Francis Bacon, René Descartes, and Isaac Newton recognized that the search for knowledge was hindered by a flawed body and a flawed mind. The body's perceptions, the workings of its sense organs, were limited and unreliable—humans can only see so far or so small and their eyes can be tricked, for example. The mind's workings could also be derailed, such as by unfortunate habits of mind or by desire. "Thoughts, for Bacon, are shaped by ways of living," John E. Leary observes.[2] But the revolution's thinkers also held that such flaws could be compensated for. With the right tools such as a telescope or a microscope, and the right method such as an experiment, the self could achieve real knowledge. In other words, the philosophical revolution depended on a self that is capable of ratiocination detached from internal and external influences, capable of observing and manipulating nature to penetrate nature's workings, and consequently capable of unraveling mysteries of God's cosmos that had hitherto remained impenetrable.

This idea of human capacity underpinned the methodological innovations of the seventeenth and eighteenth centuries, drove the creation of institutions such as the Royal Society in England, and provided the foundation for the epistemology that suffused the modern Western world. As I discuss in the following chapters, it is also an idea that attracted the attention of early novelists, who engaged with these notions as they created the novel. These concepts neither emerged fully formed from philosophers' brows in the seventeenth century nor achieved immediate acceptance within and beyond philosophy. The discussion that ensued occupied minds from a variety of backgrounds and occupations and with a variety of agendas and views, and it occupied those thinkers and writers for decades if not longer.[3] This discussion also, however, refined those ideas about the self and consolidated its place in English culture. During the period that occupies this study, that process of refinement and consolidation was still very much underway. It is a period characterized by an engagement with aspects of the self posited by natural philosophy rather than an acceptance of that self.

The Stable Self

There are several key qualities of this notion of the self that concerned not just natural philosophers but also the many people beyond philosophy who debated the self and other developments of the philosophical revolution. One key quality is its stability and unity. For natural philosophers like Bacon, there was a "there" there, an entity—he called it "the human understanding"—that is the amalgam of the body and the mind. Its workings are affected by the workings of those two aspects of human existence as well as by the dynamic of their interaction. "The human understanding is of its own nature prone to suppose the existence of more order and regularity in the world than it finds," Bacon writes, or "The human understanding when it has once adopted an opinion (either as being the received opinion of as being agreeable to itself) draws all things else to support and agree with it." He characterizes its functioning as an "unquiet" but consistent whole: "The human understanding is moved by those things most which strike and enter the mind simultaneously and suddenly, and so fill the imagination; and then it feigns and supposes all other things to be somehow, though it cannot see how, similar to those few things by which it is surrounded."[4] The self is malleable, susceptible, and created by the workings of the body and the mind, but there is only one self and it stays with a person.

Although the body and the mind were both components of this self, their relative importance differed among different schools of thought. René Descartes was able to dismiss the body, for example, while Robert Boyle found it both indispensable and ineluctable. In

England, there was diversity of thought on the roles and importance of each, not to mention debates about the role of a third element, the soul. Those debates directly manifested as disagreement about whether humans had a soul, where it was, and what it did.[5] They also appeared more indirectly, such as in the period's anxieties about ghosts and other supernatural phenomena, as I discuss in chapter 3, or in hybrid analyses of human biology. Thomas Willis, meticulously explaining epilepsy in recognizably neurological terms, connected body and soul, arguing that the "health of the Soul, should take its beginning from the restored health of the body."[6] Early natural philosophers such as Boyle, Hooke, or Newton also were sometimes somewhat disingenuous when it came to the role of the body. For these sons of Bacon, the senses were highly unreliable and the mind highly influentiable, which should have rendered highly suspect anything involving the body. Certainly, the technophilia of the late seventeenth century driving the development of ever-more-precise microscopes, telescopes, air pumps, watches, and the like was the result of this suspicion of the body and its perceptions. Nevertheless, early experimental philosophers seemed to have no concerns about using their own bodies to conduct experiments, nor did they hesitate to view their own particular bodies in general terms.[7] Under a variety of pressures on epistemology over time, the body's contribution lessened and the mind's increased. Misty Anderson notes that although Lockean and Cartesian thought were considerably different in many respects, for example, they shared an emphasis on the mind over the body.[8]

Overall, whatever their relative importance, thinkers generally viewed the self, with or without a soul, as a fusion of body and mind distinguishable and detachable from its context, and operating through a combination of experience and cogitation. The body supplied perception and experience, the mind processed those perceptions and experiences, and together the body and the mind made a self capable of seeking accurate knowledge about nature, God's creation. In *The Christian Virtuoso* (1690), Robert Boyle contends that the "consideration of the Vastness, Beauty, and Regular Motions" of nature "may justly induce him, as a Rational Creature, to Conclude" that there is God, "And this is strongly confirm'd by Experience."[9] Describing the method and the man, Boyle writes, "For they consult Experience both frequently and heedfully; and, not content with the Phenomena that Nature spontaneously affords them, they are solicitous, when they find it needful, to enlarge their Experience by Tryals purposely devis'd; and ever and anon Reflecting upon it, they are careful to Conform their opinions to it; or, if there be just cause, Reform their

Opinions by it."[10] While the knowledge may change and therefore belief may change, nevertheless, the self remains stable. Similarly in the Preface to *Micrographia* (1665), Robert Hooke contends that "by Rectifying the operations of the Sense, the Memory, and Reason," knowledge will be established and "our command over things is to be establisht."[11] Under all of these methods, tools, and claims lies the idea that there is a consistent self capable of ignoring, eradicating, or compensating for flaws both internal and external. Boyle's dedication to defining the ideal experimenter such as in his *Christian Virtuoso* testifies to the centrality of the experimenter's self in the new epistemology, but the person composed of experience and mind would return, overtly theorized or implicitly described, throughout the seventeenth century and into the next. This is what Dror Wahrman calls a "well-defined, stable, unique, centered self."[12] Natural philosophy's ideas about the unified, detached self contributed significantly to larger notions of what Timothy Reiss calls "who-ness" that were developing during this period.[13]

These ideas about the contributions of the body and the mind did not support the inclusion of women in natural philosophy or in notions of the self. Until recently, few scholars have considered the role of gender in constituting the new philosopher and the new self. In their seminal discussion of early science, *Leviathan and the Air Pump*, Steven Shapin and Simon Shaffer leave intact the assumption that men were the arbiters of knowledge and later studies such as Shapin's *Social History of Truth* and Philip Carter's *Men and the Emergence of Polite Society, Britain, 1660–1800* continue the trend.[14] For masculinity to take on this definition and function, however, ideas of gender and the self had to develop in concert. As Donna Haraway puts it, "It was the general absence, not the occasional presence, of women of whatever class or lineage/color—and the historically specific ways that the semiotics and psychodynamics of sexual difference worked—that gendered the experimental way of life in a particular way."[15] Boyle, whose *Christian Virtuoso* defined for many the ideal new philosopher, "worked to produce a new form of masculinity conducive to the new science as he envisioned it," Elizabeth Potter explains.[16] The natural philosopher was male and the self was male, and that maleness was vital for the epistemology that enabled and was enabled by this notion of the self.

Readers at the time recognized that the self being conceptualized and empowered was male, as well.[17] "In this culture there was no legitimacy in giving political voice to those who could not be presumed to speak truth, and who were, in the event, spoken for by others," Steven Shapin explains.[18] As early as the 1660s, Margaret Cavendish, Duchess of Newcastle observed that the epistemology introduced through natural philosophy had social and political implications, particularly for women.[19] Aphra Behn and Eliza Haywood also used their narratives to explore and expose the implications of such ideas for men, women, and society. Definitions of the "individual" that came to underpin English political theory and the ideologies and structures built upon it subsumed and absorbed "woman."[20] Erica Harth points out that when the universal rational subject is supposedly sex-neutral that subject is really male, thus simultaneously erasing and excluding women from notions of identity and selfhood.[21] Joan Wallach Scott's formulation—"The abstract rights-bearing individual who came into being as the focus of liberal political debate in the seventeenth and eighteenth centuries somehow became embodied in the male form"—contains two key points: "somehow" is the ascendance of an epistemology being developed through the philosophical revolution, and embodiment or the material body played a crucial role in this ascendance.[22]

There was little room in such a view for what Scott calls "feminine particularity" except to "secure[...] the universality of the masculine representation."[23] Furthermore, even among thinkers who granted a role to the body in the production, acquisition, and processing of knowledge, the woman's body was problematic. Elizabeth Potter points out that in defining the ideal philosopher, Boyle and the Royal Society not only gendered that philosopher and philosophical spaces male but did so by defining and excluding the female body.[24] Whether a female body could contain or produce rational thought, how much and for how long, remained questions throughout the period. Mary Astell thought a woman's body could; *The Spectator*'s authors were less sure; a century later Mary Wollstonecraft still had to defend the proposition. As Laura Brown shows, much late-seventeenth- and eighteenth-century misogynist writing attempted to extract the female body from Lockean notions of autonomy and self-possession.[25] Thomas Laqueur's narrative of medicine's shift from a single-body to a two-body system in defining gender has called attention to the period's changing understanding of the body, and scholars such as Karen Harvey and Laura Gowing have usefully complicated and modified Laqueur's account.[26] From the beginning of the philosophical revolution, women found their opportunities—literal or figurative, experimental or intellectual, social or individual—bound up in notions of what the body, or a woman's body, could do.

A key moment in the discussion of the self occurred at the end of the 1680s and the beginning of the 1690s with the publication of two works: Isaac Newton's *Philosophiae Naturalis Principia Mathematica* (1687), usually called the *Principia*, and the second edition of John Locke's *Essay on Human Understanding* (1694). Both men grappled with the relationship between the intellectual faculties and the body's perceptions in the generation of knowledge, albeit in different ways, and both found satisfactory solutions in detaching, either theoretically or methodologically, the body and the mind. Newton's Preface to the Reader in the first edition of the *Principia* begins by making the body's work—"Mechanicks"—the foundation of the mind's work— "Geometry." Newton writes,

> For the descriptions of straight lines and circles, on which Geometry is founded, belongs to Mechanics. Geometry does not teach the drawing of these lines, but requires it. For it requires that the beginner learn to describe these [lines] accurately before he reaches the threshold of Geometry; then it teaches how problems may be solved through these operations. To describe straight lines and circles are problems, but not geometric ones. The solution of these is required from Mechanics; the use of the solutions is taught in Geometry. And Geometry prides itself that from so few principles brought from without, it accomplishes so many things. Therefore Geometry is founded on mechanical practice and is nothing else than that part of universal Mechanics that proposes and demonstrates the art of measuring accurately.[27]

The *Principia* starts with two sections of mathematics on which the third section is built, however, privileging "ratiocination" over the work of the senses. Newton adds near the end of the Preface, "*But I hope that the principles set down here will shed some light either on this method of philosophy or on some truer method,*" signaling his interest in establishing the truth of his conclusions and equally important, the truth of his method.[28] The *Principia*'s synthesis of mathematics and experiment enabled Newton to balance or counteract the role of perception in the generation of knowledge.[29] "This geometry provided the means to move from the concrete experiences of the experimenter to the abstract hypotheses of the speculator," Judith Zinsser explains, "and thus to create mathematical idealizations of the real world."[30]

Locke performed a similar maneuver in the second edition of the *Essay*, where he claimed that the body provides experience, the mind creates ideas, and together they supply "our present sensations and perceptions: and by this everyone is to himself, that which he calls *self.*[31] Locke's point that consciousness and the body could be detached, so that a person could be himself even if infused into a different body, effectively rendered all bodies the same and all minds different, and separated the latter from the former when it came to knowing (II.xxvii.11–7). Newton and Locke appear to have reached for related solutions to the problems of the fraught connection between the body and the mind: both men used their writing to construct and propound, either implicitly or explicitly, the idea that the mental faculties could be disentangled and could function apart from the body's faculties.[32]

It is tempting to view Newton and Locke's contributions as unique, watershed events. Newton's contemporaries recognized that Newton had shattered old notions of the cosmos and provided new notions in their place. Introducing the *Principia*, Edmund Halley proclaimed that "Questions that tormented the minds of ancient scholars so often, / And even now trouble our scholars in vain with raucous strife, / We see clearly, learning routing the cloud of ignorance."[33] Whatever the reassessment of Newton has been over the centuries, there is still agreement that everything changed following the *Principia*. Similarly, Locke's ideas in the *Essay Concerning Human Understanding* about perception, consciousness, and the separability of subjectivity and the body opened new possibilities for social and political identity, relations, institutions, and systems. Early in his *Remarks upon an Essay concerning Humane Understanding* (1697), Thomas Burnet suggests that Locke's ideas of the connection between the body and morality better serve "Politicians" than "Divines and Philosophers."[34] He had a point. It would be as difficult to imagine the current global political map without Locke as it would be to imagine science today without Newton.

Nevertheless, the ideas of the *Principia*, the *Opticks*, and the *Essay* came into being in response to other ideas in circulation at the same time. Newton's math itself might have been new but other aspects of his methodology were consistent philosophically and practically with approaches used by some Fellows of the Royal Society and other natural philosophers of the day.[35] In works like *Another Essay in Political Arithmetick* (1682), William Petty, a founder of the Royal Society, demonstrated how mathematics could be used to analyze and improve the human condition.[36] A reviewer in *Philosophical Transactions* observed that such work "made it appear that Mathematical Reasoning, is not only applicable to Lines and Numbers, but affords the best means of Judging in all the concerns of humane Life."[37] In the

same issue, the editor of *Philosophical Transactions* printed translations of mathematical treatises "to manifest by Experience, the great
use of Mathematical Knowledge."[38] Philosophers had been using
mathematics to support and explain observation and experiment
since Galileo, and there were practitioners of this method in England
before and during Newton's time, not simply after.

The same is true for Locke: his method, many of his ideas and questions, and not a small portion of his rhetoric in the *Essay* were supplied by the discourse in which he participated. Locke incorporated
and rejected ideas, methods, and rhetoric from Descartes, Thomas
Hobbes, and Henry More. A great deal of Locke's material was influenced not just by natural philosophy, but by the natural philosophy
propounded by Fellows of the Royal Society, especially Boyle and
Newton.[39] "The commonwealth of learning, is not at this time without master-builders, whose mighty designs, in advancing the sciences,
will leave lasting monuments to the admiration of posterity," Locke
writes in the Epistle to the Reader that opens the *Essay*, identifying
"Boyle" and "Sydenham" as "master-builders" and "Huygenius, and
the incomparable Mr. Newton" as "the great."[40] "The *Essay* brings
to the mind and its 'ideas' the methodology of rigorous observation
championed by the Royal Society, but also a rhetoric of diffidence and
cautious consensus practiced by so many of Locke's scientific contemporaries," Peter Walmsley points out, arguing that "Locke's rhetorical context" includes the *Essay*'s "social and intellectual moment."[41]
For all their individual genius, the *Principia*, the *Opticks*, and the
second edition of the *Essay Concerning Human Understanding* functioned like precipitates: elements forced out of a solution that make
visible both the fact that the solution contains disparate elements and
that it contains a particular element. Concerns about the fallibility of
the human being, about the roles of perception and reason in making
knowledge, about the limits of the human capacity to know nature,
were all in circulation. These works are particularly brilliant instances
and articulations of those concerns.

Equally important, these works also functioned as catalysts: sub-
stances whose addition to a solution causes a reaction that reveals
something new. Although neither man's work immediately gained
acceptance and neither did the concept of the self that underpinned
the epistemology that shaped their work and that was in turn shaped
by it, reaction set in. Decades of debate ensued, consuming political,
theological, scientific, economic, and social thought. Locke's move
to redefine the self from the theological definition "soul + body" to
what Amos Funkenstein would call a "secular theological" definition

of "mind + body" took years to accomplish, even when that process was augmented by public debates such as the one between Samuel Clarke and Anthony Collins in 1707–8.[42] Similarly, although the *Principia* became a tremendous success in philosophical circles, cultural as well as philosophical success for Newton and his method required the combined impact of the Boyle Lectures and the publication of the *Opticks* (1704).[43] As Marjorie Hope Nicolson points out, Newton really entered the mainstream poetic imagination after 1727, the same period in which he began to achieve "apotheosis," in Mordechai Feingold's term.[44] Catherine Trotter Cockburn and Winch Holdsworth dueled over Locke's work in the 1720s; Joseph Butler was attacking Locke in the 1730s; Vincent Perronet was defending him.[45] By 1730, Newton's and Locke's ideas might have become culturally compelling, but they had not yet become fully culturally accepted. Not everyone thought they were right; not everything was written within their epistemology.

Detachment and Morality

Another issue of the new notion of the self was its relationship with context. There was general agreement among revolutionaries like Bacon, Boyle, and Locke that the self could be detached from its context and that indeed, it existed autonomously. As Eve Keller summarizes, "Viewing nature as a discretely ordered and predictable machine permits a commensurate view of the self as an autonomous knower whose subjectivity need not enter into the evaluation of knowledge production—a self we now loosely call 'the humanist individual.'"[46] This was a new idea and even a strange one, but it was crucial to new notions of the self.[47] If the self could not be detached from its context and therefore its influences, it had no hope of achieving true knowledge. This separation operated on different levels but was equally significant on each. There was the idea that the self stood apart from the rest of creation and could therefore use its position to study nature, a view perhaps epitomized by the experimental approach to knowledge-gathering. There was also the related idea that the self could strip away the encrustations of its context in order to perceive and reason accurately, a view epitomized by the Cartesian dependence on cogitation and what we might call theorizing. In either case, however, the idea of separation had a price. Without a connection to others and to the educational forces of society, the self stood in considerable danger of forming dangerous, even immoral ideas, for example. It also stood in danger of not forming ideas that enabled an individual to navigate society. While that might be problematic for the individual, it also had serious and larger consequences. How could civilization hope to continue if its members did not know how to be civilized? Part of the debates about the nature of the self therefore had to do with the problem of detachment. In natural philosophy, "detachment" was both a quality and a literal situatedness that was necessary for pursuing and acquiring knowledge, but also seemed dangerously liberating. In *The New Organon*, Bacon carefully distinguishes the four Idols from true faith. The Idols—received, conventional wisdom (the Idols of the Tribe); individual perception and thinking (Idols of the Cave); the distortions of common language (Idols of the Marketplace); and systems of thought already in place (Idols of the Theater)—make it impossible to achieve true knowledge. Although these notions must be abandoned, Bacon is careful to note that not all notions should be. "There is a great difference between the Idols of the human mind and the Ideas of the divine," he writes: one should keep the latter and remove the former.[48] Descartes begins his famous project by wiping his own slate clean, with the exception of faith. "As regards all the opinions to which I had until now given credence, I could not do better than to try to get rid of them once and for all, in order to replace them later on, either with other ones that are better, or even with the same ones once I had reconciled them to the norms of reason," he explains in the *Discourse on Method* (1637), but he also says that in order to pursue this plan, he "formulated a provisional code of morals" that started with obedience to the "laws and the customs of my country, [and] constantly holding on to the religion in which, by God's grace, I had been instructed from my childhood."[49] From its early days, then, one strain of thought in the philosophical revolution required both the removal of preconceptions and the preservation of some kind of code. While there was some disagreement during the seventeenth and eighteenth centuries about whether those principles were or should be religious or secular, the larger issue was the relationship between a philosopher's values and his method.[50] Despite the need to eliminate misprisions, there was still an interest in using standards for right and wrong to help guide the new breed of philosopher.

Underpinning these ideas is the concept that the self can be distinguished from forces such as other people that can pervert true understanding and even true perceiving. If before the philosophical revolution, the sense of self was defined by elements such as "material world, society, family, animal being, rational mind, divine" then the self, defined by these "spheres" to use the classical phrase, was essential. You "were" these things, Reiss explains: "At the same time they were public and collective, common to everyone *qua* human."[51] By the late seventeenth century, however, some of these things—the material world, society, and family, for instance—were considered in some circles as entirely separate from the self and vice versa. Although separate, they were not equal. During the late seventeenth and early eighteenth centuries, those external forces and entities were understood to shape the self and its perceptions. Perception and whatever made it thus had to be evaluated carefully and rationally.[52]

Consequently, if the philosopher needed to strip away erroneous beliefs and could do so because his self was distinct from other selves and forces in nature, then the philosopher should be detached: from his data, from belief and desire and opinion, from ego, even sometimes detached from society. Detachment, as Peter Dear and Peter Harrison have pointed out, thus became a fundamental characteristic of the philosophical revolution's methods over the course of the seventeenth century.[53] Many Fellows of the Royal Society saw a connection between denuding oneself of company and of the influences of society. In *The Christian Virtuoso*, Boyle evokes the image of a swarm of pernicious "Atheists" and "Ingenious Men" whose "contagious Company must Endanger, if not Infect, me."[54] Similarly, in her poem "An Invitation to my Friends at Cambridge" (1688), Jane Barker invites the learned students to join her in "Solitude," where "we have full enlargement of the mind" so they can continue their studies.[55] Even if physical and intellectual isolation was more a pose than a practice, nevertheless the idea that detachment or disengagement was crucial to successful philosophical endeavor remained.[56]

This relationship between the individual and separation had a profound impact on the values and epistemology developing during this period. As Lorraine Code points out, in this system, "Knowers are detached, neutral spectators, and the objects of knowledge are separate from them."[57] A premium on detachment underpinned the drive to establish mathematics as a method for investigating and predicting the behavior of the cosmos. Although mathematically based natural philosophy did not control natural philosophy during the period between 1660 and 1730, it surged in importance after Newton published the *Principia* in 1687 and again after 1704, once Newton became president of the Royal Society and his deification was in full swing.[58] This approach made its way into the cultural mainstream, as well. As Jesse Molesworth argues, the "widespread encroachment of mathematical thinking" created a simultaneous reaction against it that ultimately produced the realist novel, a form which, as Molesworth notes, depends heavily on a set of entirely unrealistic and improbable events masquerading as logic and order.[59] The realist novel was not everyone's novel and it was certainly not the amatory novel, which dominated the 1720s.[60] As I discuss in later chapters, however, other types of narratives were indeed capable of responding to the rise of mathematical epistemology.

Although detachment was gaining epistemological value, philosophy also recognized a role for the community. That role was primarily methodological. The philosopher was tethered to others by the need for a community of "modest witnesses," other people who could validate his method and conclusions by watching how the method was applied and the conclusion reached. "Matters of fact were the outcome of the process of having an empirical experience, warranting it to oneself, and assuring others that grounds for their belief were adequate. In that process a multiplication of the witnessing experience was fundamental," Shapin and Shaffer explain.[61] The modest witness or modest witnesses were thus another hedge around the fallibility of the self. As early as Francis Bacon and William Harvey, English philosophers advocated a role for a group of like-minded, properly-trained colleagues to ensure that each individual and philosophers overall produced only true knowledge. In this schema, the self's discoveries did not constitute knowledge until the process could be replicated and the conclusions reached by other selves. "I was greatly afraid lest I might be charged with presumption...unless I had first proposed the subject to you, had confirmed its conclusions by ocular demonstrations in your presence, had replied to your doubts and objections, and secured the assent and support of our distinguished President," Harvey explains in *On the Motion of the Heart and Blood in Animals* (1628).[62] There was no difference between the philosopher and the modest witness except in terms of function at any given moment. The modest witness's job was to validate or reject the findings of another individual self. The relationship between the individual and the group thus establishes a kind of compound self, a many-headed self as it were, whose combined perceptions and conclusions created knowledge. This aspect of seventeenth- and eighteenth-century natural philosophy interested writers like Jane Barker, who recognized that an epistemology built on communal knowing offered all individuals, even those disenfranchised by religion or class or gender, an opportunity to participate in the identification and dissemination of knowledge.

For one school of English natural philosophers, this connection between individual and group is what Steven Shapin calls the "moral bond between the individual and other members of the community."[63] By the middle of the eighteenth century Colin Maclaurin could write, "We are now arrived at that happy stage of experimental philosophy...when not private men only, but societies of men, with united zeal, ingenuity and industry, prosecuted their inquiries into the secrets of nature, devoted to no sect or system."[64] The rise of philosophical groups in England during the eighteenth century testifies to the perceived relationship between sociability and natural philosophy, and to the eventual dominance of this strain of thought.[65] Other philosophers such as Thomas Hobbes saw this connection and its valorization as more problematic, however. While the bond between individual and group involved a code of ethics that permitted, if not required, the honest sharing and evaluation of information and insight, it also applied only to what Shapin and Shaffer call the "relevant community," members of the philosophical community, such as among Fellows of the Royal Society.[66] When Frederick Slare published *Experiments and Observations upon Oriental and Other Bezoar-Stones* (1715), for example, he took care in the title and Dedication to proclaim not only that he was a Fellow in the Royal Society but also that its members had heard and approved his paper on that topic.[67] Similarly, although Boylean experimentalists opposed Newtonian mathematicians on the grounds that mathematically based philosophy excluded all but the few people who could understand it, they were concerned that it limited the philosophical community in which knowledge circulated and was validated, not that it restricted ordinary people from understanding and participating.[68] In detaching the self from its context, natural philosophy raised questions about the self's relationship with the larger community, including the definition or parameters of the community and the nature of the relationship.

As the language around the bond between individual and group suggests, philosophers also attempted to counter the potentially amoral detachment of the philosopher by infusing natural philosophy with a moral component. Locke believed that his model of the self, with identity detachable from the body, supported morality although his opponents Edward Stillingfleet and Thomas Burnet argued just the opposite, sometimes vehemently. Boyle asserted that no discipline "does more enable a willing mind to exercise a Goodnesse beneficiall to others" than natural philosophy.[69] Justifications like this one continued to focus on the benefit to the philosopher, however. More often, experimental philosophers as well as their opponents focused on "use value" for infusing their method and conclusions with moral value.[70] Robert Hooke's preface to *Micrographia* claimed that the Fellows "principally aim at such, whose Applications with improve and facilitate the present way of Manual Arts."[71] Addressing the other Fellows in 1725, Thomas Morgan wrote, "You have the Health and Ease, the Lives and Happiness of your fellow Creatures too near at heart, than to sacrifice those valuable Interests to any darling Hypothesis, indisputably receiv'd, and establish'd into a System of Physical Orthodoxy."[72] In the 1720s, Aaron Hill and Daniel Defoe propounded the Whiggish idea that knowledge should be pursued for how it can better humanity's condition. In his *Account of the Rise and Progress of the Beech-Oil Invention* (1715), Hill announces, "I should be glad, if the Example could stir up in Merchants and Tradesmen of all orders, a vigorous Application to Searches after new Discoveries in their respective Callings. Such an universal Bent of Genius would not only enrich a thousand private Families, but improve our Trade, encrease our Wealth, and firmly establish the Glory of the Nation."[73] Christine Gerrard notes that "scientists" and "projectors" were linked practically and in the public imagination because the discoveries of one made the projects of the other possible.[74] In that regard, a relationship might be said to exist. The same held for the philosophy that was accessible to women. In advocating the benefits of natural philosophy for women in Eliza Haywood's *The Female Spectator* (1744–46), "Philo-Naturae" sounds remarkably Boylean, stating that the study of nature, complete with magnifying glass, provides "not only a most pleasing Amusement, but is the best Lesson of Instruction we can read, whether it be applied to the Improvement of our Divine or Moral Virtues," and it lubricates social interactions.[75] The self might be intrinsically detached but philosophy countered the potential entropy and chaos of a universe of unattached selves by valuing actions that established a bond with others and that contributed to the well-being of the group.

As philosophers' efforts to justify this view of the self suggest, this intersection of ideas about intellectual isolation and social isolation, of principles and a lack of influences, generated considerable anxiety. Did the self really exist as a discrete entity that established connections to others at will, rather than inevitably or by nature? What was the relationship, if any, between such an autonomous self and its social context in the production of valid knowledge? In the early decades of the eighteenth century, in fact, the relationship between the conditions of knowledge production and morality was in flux. Reiss points out that the "idea of a self free and independent in its will, intentions and choices;...of a separate, private individual" was still an odd one at the end of the seventeenth century.[76] It took far longer for the "significance of the moral character of the agent of investigation" to become essentially "irrelevant," as Harrison puts it.[77] As writers like Aphra Behn, Jane Barker, Eliza Haywood, and Mary Davys recognized, this idea of the self had consequences for the ways in which that self was represented in and conceptualized by language.

Like questions about the role of the body, questions about the relationship between the self and the community and thus between detachment and morality had a particular resonance for women. Women were expected to view themselves as constructed by and contributing to the group, obedient to parents, aspiring to marriage, and responsible for children. The pressures on women to conform to this notion of virtue were considerable, to put it mildly. Autonomous action and independent perception and analysis were not part of conventional femininity. Chastity, civility, compassion, generosity to the poor, obedience, and so on, all behaviors tying the female individual to her community, were. A woman was also expected to be a civilizing and moralizing force, responsible for training boys to be part of the society that privileged them as men.[78] Richard Allestree, in his popular conduct book *The Ladies Calling* (1673), asserts that "all Mankind is the Pupil and Disciple of Female Institution: the Daughters till they write Women, and the Sons till the first seven years bepast [sic]; the time when the mind is most ductile, and prepar'd to receive impression, being wholly in the Care and Conduct of the Mother."[79] *The Ladies Calling* remained in print until its eleventh printing in 1720. A woman was expected to see her self as woven into the fabric of a group, constituted by it and constituting it in turn. This special moral role was not entirely empowering, of course. It was often used to marginalize or justify the marginalization of women. Her body, for example, rendered her inferior to men, regardless of how writers might try to make that inferiority appealing.[80] Furthermore, if epistemology valorized the isolated individual (the man) then the individual who could not or ought not exist as an isolated entity (the woman) was removed from the systems of knowledge production.[81] Women writers could be very interested in the moral and social implications of detachment as a value, not only in terms of the consequences for women—although as Haywood's story-within-a-story in *The Tea-Table* shows, authors recognized the stakes for women. Authors like Haywood and Mary Davys also recognized the possibilities, both positive and negative, that detachment offered for moral knowledge and moral credibility. Their novels explore both the benefits and the costs of being separate from others and from everything.

Writing the Self

Thus Haywood's and Davys's novels, like Behn's and Barker's, illustrate how attractive authors of many interests and backgrounds found the need and the challenge to represent this new self in writing. Michelle M. Dowd and Julie A. Eckerle observe that for authors, "categories of genre, gender, and identity were thus mutually constitutive in early modern England" from at least the late sixteenth century through to the beginning of the eighteenth.[82] Natural philosophy's problems of language were far-reaching and significant socially, politically, and literarily. For Bacon, the relationship between language and what it described or represented was troubled. Bacon's heirs in England and particularly in the Royal Society struggled to develop a rhetoric that would represent stable truths without being destabilized by an imaginative component.[83] The "tradition of seventeenth-century language reform" traces back at least as far as Francis Bacon. A "reformed, rational language" concerned the Oxford group of the 1650s, the precursor of the Royal Society whose membership included John Wilkins, who wrote *Essay towards a Real Character and a Philosophical Language* (1668), Thomas Willis, Robert Boyle, Christopher Wren, John Locke, John Wallis, and Robert Hooke.[84]

Innovations to language and concomitant innovations to literary form reveal the inadequacies of extant modes of writing for the new notions of the self and knowledge. For men writing about the male self, it was a profoundly difficult task to conjure a sense of the self in writing. The struggles of an author like Robert Boyle, who was not only prolific but also various in his writing, reveal how crucial this coherent, representable self was to the philosophical revolution.[85] Shapin and Shaffer describe the development of "literary technologies" for establishing the author as a "provider of reliable testimony," as "such a man as should be believed."[86] Jan Golinski points out that Boyle roamed among different genres in an effort to find the one best suited to representing the self, the workings of the self, and the knowledge attained through the workings of that stable, unified self.[87] For women writing about their experience and insights, the problem of rhetoric was even more acute as both established modes and the new efforts by men were often inadequate or inappropriate to women's purposes.[88] Like Boyle, Cavendish also struggled to find a genre in which to express and represent the self, employing established genres, adjusting their conventions, and creating her own literary forms as her needs required. As Eve Keller points out, *Observations on*

Experimental Philosophy and *Blazing World* are companion pieces, presenting the same challenge to experimental philosophy in different forms as Cavendish sought the right form for the job.[89] In general, in fact, texts discussing natural philosophy used a wide variety of genres including dialogue, narrative, and essay. The pages of *Philosophical Transactions*, which carried the questions and conclusions of Europe's natural philosophers to each other, reveal a variety of genres, some used within the same report, for the circulation of new ideas and methodology.

One key to establishing credible authority in a text was the representation of the knower and his knowing. Shapin and Shaffer point out the necessity for strategies that would establish the author as a "man of good faith."[90] These strategies often centered on the representation of the man himself. "The man who wrote up an experiment for the reading public had to be trusted," Elizabeth Potter notes.[91] By the late seventeenth century, letters were a well-established mode for emphasizing the presence and authority of the writer, whether actual or fictional. Epistles were familiar to men and women as public, private, and pseudo-private productions. In the late seventeenth century letters came in a variety of subgenres: familiar letters, what Amy Elizabeth Smith calls "formal" letters, statesmen's letters, royalist letters, and the like.[92] Men and women both carried on correspondence and letters were regarded as a way of demonstrating the character of their author. There were handbooks on writing letters.[93] Samuel Hartlib and his successor, Henry Oldenburg, used letters to foster a community of natural philosophers but even without their encouragement, natural philosophers across Europe used correspondence to share information, findings, and queries. Because letters, as Gerald MacLean explains, "lay claims to ownership and authenticity: to material objects or particular experiences, to eyewitness knowledge about events or newly revealed truths, to specific rights to do something or go somewhere," they were particularly useful for establishing authority and authenticity, fact and knowledge.

Furthermore, as MacLean notes, one of those objects to which letters lay claim is the identity of their author.[94] Two natural philosophers profoundly concerned with establishing their ideas and their selves, Cavendish and Boyle, for example, used epistolary form in several philosophical publications.[95] Boyle's declaration of the experimenter's self, *The Christian Virtuoso*, opens as a letter and consistently addresses the correspondent.[96] According to Dwight Atkinson, slightly more than half of the reports in the *Philosophical Transactions* in 1675 took the form of epistles to the Royal Society.[97]

"To an overwhelming extent, the Society *was* its correspondence" at this period, Roger Iliffe explains.[98] A 1665 report of chronometers, for example, includes an excerpt from a letter by Christian Huygens, who built the clocks and who writes in first person.[99] Philosophers far less established, such as the anonymous author of *Two Essays Sent in a Letter from Oxford, to a Nobleman in London* (1695), also used epistolary form, suggesting that it was a mode widely used by those wishing to join the discourse.[100] Letters were crucial to natural philosophy's publications, both for how they enabled truth claims about their contents and for how they enabled a truth claim about the self that they presented.

Letters were not always up to the task, however. Another technique for representing the knower and authoritative knowing was narrative. The first-person point of view, so helpful in creating the image or sense of stable, unified self in letters, also proved useful in narrative forms. Recounting his method and conclusions in *On the Motion of the Heart and Blood in Animals*, for example, Harvey emphasizes the presence of the experimenter. "Experimenting with a pigeon on one occasion," he reports,

> ...After the heart had wholly ceased to pulsate, and the auricles too had become motionless, I kept my finger wetted with saliva and warm for a short time upon the heart, and observed that under the influence of this fomentation it recovered new strength and life, so that both ventricles and auricles pulsated, contracting and relaxing alternately, recalled as it were from death to life.[101]

Here Harvey's description of the workings of the pigeon's body includes allusions to his own body. Similarly, Robert Hooke depends on the first-person singular from the opening of *Micrographia* (1665), using the interjected commentaries to create a persona of the ideal, ideally acceptable scientist, "as undogmatic, as a Baconian, as an orthodox Christian, and as a follower of the mechanical philosophy."[102] Hooke refers to his "sincere hand and faithful eye" in describing the method and findings presented in *Micrographia* and like Harvey, used the active voice to convert the *Micrographia*'s "voice" into the sense of a being who was really doing the experiments.[103] This approach was widely used in the Royal Society's keystone publication, its journal *Philosophical Transactions*, whose reports from throughout Europe and around the colonized world depended heavily on "author-centered" writing and the idea of coherent, stable self. Through a variety of strategies, its reports provided a sense of an author who had collected the data, done the observations, performed the experiments, traveled to the particular place, and so on.[104]

Even when natural philosophical writing seemed to attempt a third-person disembodiedness more familiar to today's readers of science writing, it often erupted into a resolute evocation of a person. The 1665 report in the *Philosophical Transactions* on the success of a new chronometer in establishing location at sea provides information that originates with "Major Holmes," who led the expedition, but does not speak in Major Holmes' voice. It opens with, "The Relation lately made by Major *Holmes*, concerning the success of the *Pendulum-watches* at sea (two whereof were committed to his care and observation in his last voyage to *Guiny* by some of our eminent virtuosi, and Grand Promoters of Navigation), is as followeth," a seemingly detached, third-person statement that breaks into first person with the pronoun "our" and the evaluative adjectives "eminent" and "grand." Similarly, the second half of the report is an extract of a letter in the first person by Huygens, who built the clocks tested by Major Holmes.[105] Hence, although the report disembodies Major Holmes, it also grants a presence to Huygens and even the unnamed author of the first half of the report (presumably Henry Oldenburg). Antonie van Leeuwenhoek's two-part report in 1675 on his work with the optic nerve also breaks into the first person. In these accounts, Leeuwenhoek provides a detailed narrative of his actions, characterized by a careful description of the steps he took and accompanied by illustrations in some cases. He also describes his thought processes during the series of experiments: "I here thought with myself," he writes, and elsewhere describes a whole imaginative process with a glass of water:

> But I imagined it might be performed, for example, after this manner; *viz.* I represent to my self a tall Beer-glass full of Water: This Glass I imagine to be one of the firmaments of the *Optic Nerve*, and the water in the Glass to be the globuls of which the filaments of that Nerve are made up, and then, the water in the Glass being toucht on its surface with the finger, that to this contact did resemble the action of a visible object upon the Eye, whereby the outermost globuls of the fibers in the Optic Nerve next to the Eye are toucht.[106]

His phrase, "I imagined to myself" recurs throughout the long description of his process, well beyond this excerpt. The complete removal of any sense of a person from the rhetoric had not become a desideratum.

Crucial to all these literary efforts is the lack of an accepted set of conventions or an acknowledged genre for representing the new self and new ideas about knowledge. While philosophers struggled to develop established forms for their work, the pressures on language and text opened up possibilities that other writers could explore, as well. There was a need to find a way or more than one way for representing the new stable, unified, reliable self and his relations with other selves as well as with the natural world. But while writers busily and creatively stretched extant genres and conventions to the task, other writers equally busily and creatively created new genres and conventions for the same task.

It is into this environment of ideas about the self that the authors of the following chapters wrote. They engaged with these ideas of human capacity: about the relationship between the body and the mind, about the role of detachment and the dangers of isolation, about the political, moral, and social implications of these ideas of what it means to be human. These authors were not alone. Other women as well as men engaged with ideas of the self in circulation. But these authors are particularly useful for recognizing the debates and the role of the debates in shaping structures that came to underpin the novel as a genre, because these authors recognized the problems of gender in these ideas of selfhood, thus highlighting through their narrative structures the philosophical issues and their larger implications.

Chapter Two

An Ingenious Romance: The Stable Self

Introducing the second edition of Isaac Newton's *Principia* in 1713, Roger Cotes announced that "Those who assume hypotheses as first principles of their speculations, although they afterwards proceed with the greatest accuracy from those principles, may indeed form an ingenious romance, but a romance it will still be."[1] Although it would have outraged him to hear such a claim, the argument could be made that the new natural philosophy in England, including the work in Newton's *Principia* that he was so unequivocally defending, was founded upon a hypothesis: the idea that humans possess a unified, stable self. Newton may have been correct about gravity, as Robert Hooke was correct about the appearance of a flea and Robert Boyle correct about a vacuum; they proceeded, in other words, with the greatest accuracy. But the first principle of their speculations, that human beings possess a stable self capable of rational and detached observation and cogitation leading to knowledge of the natural world, appeared to some thinkers and writers to be more hypothesis than fact. Furthermore, those thinkers and writers were often uncomfortable with the impact that this idea of self had on the definition of knowledge and the attendant distribution of power in their world.

One of those writers was Aphra Behn. Behn's interest in, or at least her engagement through writing with the revolution in natural philosophy was particularly strong in the 1680s, encompassing translation, adaptation, poetry, and prose. While her works often took on specific issues in natural philosophy, many of them were concerned with the idea of selfhood being propounded by natural philosophers of the revolution. Behn's narrators, particularly in *Love-Letters Between a Nobleman and His Sister* (1684–87) and *Oroonoko* (1689), exemplify their author's considerable, ongoing concerns with the idea of the self propounded by natural philosophers.[2] Behn's experiments in narrative during the 1680s reveal an increasingly complex investigation of and challenge to the idea of the self, the concept of knowledge that depends upon this idea, and the social and political consequences of that epistemology.

Although by the end of the seventeenth century it was not necessary to be active in philosophical circles to be aware of philosophical debates and issues, Behn's work and life intersected with the New Science on several levels, particularly during the last decade of her life. She was familiar with the work of philosophers including Lucretius and Thomas Hobbes. Her friends and lovers read the work of Robert Boyle, Hobbes, and others; she knew Thomas Sprat and Thomas Creech. Her writing reflected her knowledge and interest.[3] She translated Bernard de Fontenelle's *A Discovery of New Worlds* (1688) from French and Book VI ("On Trees") of Abraham Cowley's *Of Plants* (1689).[4] These translations do more than reproduce the original; they also critique, adjust, and respond to the ideas within and underpinning the source texts. While such revisions testify to Behn's facility with other languages, they also demonstrate Behn's familiarity with and critical thinking about the new philosophy.

Behn expressed her interest in the revolution in natural philosophy in her own original works, as well. There are offhand references such as when Lady Desbro tells Freeman that his "great Pleasure to cheat the World" is "Power, as divine *Hobs* calls it," but elsewhere Behn's treatment is more sustained and her knowledge clearly considerable.[5] Her poem to Thomas Creech celebrating his translation of Lucretius's *De Rerum Natura* (1683) reveals an understanding of Lucretian ideas, particularly her playful use of atomic theory in describing her own material composition:

> But I of feebler Seeds design'd,
> While the slow moveing Atoms strove
> With Careless Heed to Form my Mind,
> Compos'd it all of softer Love....[6]

Like Thomas Shadwell in *The Virtuoso* (1677), Behn satirized the New Science in her play *Emperor of the Moon*, produced in 1687, the same year that Newton's *Principia* was published. In it, Behn recognized and attacked the relationship between experimental philosophy, spectacle, and voyeurism, suggesting that the observing and "modest witnessing" required of English experimental philosophers such as the Fellows of the Royal Society was hardly the detached, rational exercise it was made out to be by its advocates.[7] The significance of desire's role in observation was one that Behn also explored elsewhere, particularly in *Love-Letters Between a Nobleman and His Sister*, composed during the same decade. Behn's writing thus not only reflects a familiarity with the work of philosophers from Lucretius to

Hobbes but also demonstrates an independent, critical application of those ideas in fresh contexts that was particularly intense and sustained during the 1680s.

This interest in the revolution in natural philosophy supplied content for Behn's narrative prose fiction even when natural philosophy was not the focus of the work. Barbara M. Benedict has pointed out how the philosophical revolution offered the authors of amatory fiction a legitimizing, sanitizing vocabulary that allowed them to portray even sexual curiosity as moral.[8] In Behn's work, natural philosophy, especially experimental philosophy of the kind performed by the Royal Society, shows up sometimes as a legitimating discourse and sometimes as one to be challenged. *Love-Letters*, for example, is riddled with allusions to the latest scientific knowledge: Silvia explains how the motion of her blood changes with her emotions and feels Brilljard's pulse to see if he is lying, Antonett describes a young clergyman wooing her with natural philosophy, and so on.[9] Descriptions in *Oroonoko* often temporarily shift the focus of the narrative from character and plot to botany, zoology, and other branches of natural philosophy. As Anne Bratach points out, Behn is drawing on natural philosophy's use of lists and meticulous description when the narrator identifies "little Rarities; as Marmosets, a sort of *Monkey* as big as a Rat or Weasel, but of a marvelous and delicate shape, and has Face and Hands like an Humane Creature: and *Cousheries*, a little Beast in the form and fashion of a Lion, as big as a Kitten; but so exactly made in all parts like that noble Beast, that it is it in *Miniature*."[10] Describing the "*Armadilly*," the narrator explains it is "a thing which I can liken to nothing so well as a *Rhinoceros*; 'tis all in white Armor so jointed, that it moves as well in it, as if it had nothing on; this Beast is about the bigness of a Pig of Six Weeks old" (43). George Warren's *Impartial Description of Surinam*, for example, provides frequent and extensive lists, such as the review of birds or of primates, as well as descriptions anchoring the object in the familiar, such as when he notes that deer "are much like our ordinary ones in England" or that "Hares more resemble a pig, than any other Creature that I know."[11]

It is not only the literally foreign that receives such treatment. Philosophers describing known objects through new perspectives were thrown back on old modes of expression, using comparisons and vocabularies from a variety of familiar discourses to explain and describe. In his *Experimental Philosophy* (1664), Henry Power describes the flea as it appears under the microscope: "It seems as big as a little Prawn or Shrimp, with a small head, but in it two fair eyes globular and prominent of the circumference of a spangle; in the midst of which you might (through the diaphanous Cornea) see a round blackish spot, which is the pupil or apple of the eye, beset round with a greenish glittering circle, which is the Iris (as vibressant and glorious as a Cats eye) most admirable to behold," adding "nature having armed him thus *Cap-a-pe* like a Curiazer in war."[12] Repeatedly and like other texts of the New Science, *Oroonoko* seeks the language to describe and explain the material reality of Surinam, as the narrator represents herself as returning from a new, wondrous world filled with new, wondrous fact and notion.

Behn reproduces this struggle to describe and identify and therefore to understand throughout *Oroonoko*. Unlike naming in *Robinson Crusoe*, which as Maximilian Novak contends is a successful bid to create and transform, *Oroonoko*'s fascination with the new is also a concern with linguistic limitation.[13] "Parties on all sides are faced with objects for which they have no known referent, and a vocabulary of newness and associated curiosity pervades the narrative," Emily Hodgson Anderson explains.[14] *Oroonoko* offers a number of episodes designed to highlight the failure of the old to make sense of the new, although the narrative also explores this issue at a more thematic level, as the narrator struggles to make sense of what she saw in Surinam: not just the indigenous creatures but also the behavior of all three groups, the Caribs, the slaves, and the British. The narrative reveals the problems of comprehending Oroonoko, who is both black and beautiful, African and European, royal and a slave, as well as Europeans, who claim to be civilized and yet enthusiastically dismember a prince.

But while aspects of *Oroonoko* suggest that Behn is partaking of the same challenges facing natural philosophy, other aspects suggest that her engagement with natural philosophy, particularly experimental philosophy, is more critical. Such an approach would be consistent with her other works involving natural philosophy, including her translations of Fontenelle and Cowley, as well as her poem on Creech's translation of Lucretius. As Anderson points out, *Oroonoko* offers spectacle and reenactment as a solution to the failure of language to transmit the new in comprehensible terms.[15] In this endeavor the narrative resembles a work like *Emperor of the Moon*, which both savors and criticizes natural philosophy for being spectacle.[16] Furthermore, episodes such as the numb eel indicate that Behn is probing experimental philosophy's weaknesses. Hearing that contact with a numb eel would numb a person and "deprive 'em of Sense," Oroonoko "cou'd not understand that Philosophy, that a cold Quality should be of that Nature" and he "had a great Curiosity to try whether it wou'd have the same effect on him it had on others" (46–47).[17] As a result, Oroonoko "often try'd, but in vain" to catch one, until "at last, the sought for Fish came to the Bait, as he stood angling on the Bank; and instead of throwing away the Rod, or giving it a sudden twitch out of the Water, whereby he might have caught both the *Eel*, and have dismiss'd the Rod, before it cou'd have too much Power over him; for Experiment sake, he grasp'd it but the harder" (47). Significantly, Oroonoko not only questions the credibility of the phenomenon because it is inconsistent with philosophical knowledge but also attempts to test it through experiment and on his own body. This technique was used by experimental philosophers including Isaac Newton, who inserted needles behind his éyes to test light perception; David Hartley, who tested Joanna Stephens' medicine for gallstones on his own urine; and Robert Boyle, who drank cold water to observe its effect on his body inside and out.[18] Oroonoko's experiment, however, challenges the assumption that the body is not itself a variable in the experiment.[19]

Behn's narrative prose fiction also reveals a consistent, ongoing exploration of the philosophical revolution's concepts of the self. In this regard, Behn was concerned with a primary issue in the philosophical revolution, one engaging philosophers from a wide variety of schools and approaches from the early seventeenth century. Although this interest appears throughout her narrative *oeuvre*, her tripartite *Love-Letters Between a Nobleman and His Sister* reveals most clearly and extensively how Behn's interest evolved during the 1680s and how Behn developed narrative structures to respond to that interest. Catalyzed by the real, deliciously scandalous elopement of Lady Henrietta Berkeley with her brother-in-law, Ford Lord Grey, Behn published Part 1, the epistolary narrative of Silvia and Philander, in 1684. As the drama continued over the next several years, Behn brought out a second part in 1685. This volume also used letters, but they were held together with the extensive commentary and explanations of a knowledgeable narrator. Again events continued, culminating finally in the Duke of Monmouth's disastrous rebellion and the exile or deaths of the major players; again, in 1687, Behn brought out a thinly disguised narrative, and again, she changed the form, this time eschewing epistolarity altogether in favor of her knowing, only vaguely involved narrator. Gone entirely were the letters that had structured the first part and costructured the second.

These structural changes are usually attributed to the work's composition in installments.[20] Such claims either make Behn's structural developments responses to external prompts or forces, or uncontextualized narrative techniques. In fact, the changing structure of *Love-Letters* reflects Behn's exploration of ideas about the self and about knowledge that were articulated primarily by natural philosophers, particularly (although not exclusively) those of the Royal Society. The shift from the multiple perspectives of Part 1 to the single perspective of Part 3 of *Love-Letters* reveals Behn's ongoing response through the 1680s to the idea of a coherent, stable self and the epistemology based on that idea. It was about the self being represented: for Behn, the self is not stable and it is not reliable because it cannot know itself or anything else sufficiently fully. These issues were not simply a matter of accuracy for Behn—are humans really how natural philosophers describe us? Also at stake, as Behn recognized, was a question of epistemology, of knowability—how do we know what we know?—and of the systems built on that epistemology. Behn's structural innovations in *Love-Letters* and other narrative prose fictions reveal her interrogation of the epistemological, and therefore of the significant social and political, stakes of these new ideas of selfhood.

The first part of *Love-Letters* derives from the amatory or "familiar" letter, a form of epistolarity that at first appears somewhat different from the epistolarity shaping and reflecting the discourse in natural philosophy. Amy Elizabeth Smith argues, for example, that the familiar letter used in amatory fiction is significantly different from the "formal" letter used in philosophy during this period and must be theorized differently as a result.[21] And indeed, initially and on its own, Behn's use of epistolary form in Part 1 is not necessarily innovative and not necessarily a direct response to philosophical discussions of the nature of the self. The epistolary form becomes more significant, however, when put in the context of Behn's ongoing formal innovations with this novel; that is, when viewed as a departure point for her later installments, Behn's use of epistolary form becomes something more than a canny, commercial decision.

As Behn's handling of the amatory or familiar letter suggests, there are crucial issues with letters as a genre having to do with their representation of a self that transcend the specifics of subgenres like "familiar" or "formal." During the seventeenth century, rhetorical strategies were carried from one form, including epistolary subgenres, to another in order to construct an authoritative self.[22] When Janet Todd claims that the restoration letter was performative, she points out a fact intrinsic to the genre overall, as the authors of early modern letter writing guides well knew.[23] Ultimately, Helen Wilcox observes, "As a mode of self-expression, therefore, the early modern letter can be richly expressive but also deceptive and ambiguous, offering and yet denying self-construction."[24] Consequently, although letters come in many kinds and serve many purposes, and may even construct different audiences and relationships between actual and authorial/ intended audience, Behn's letters in Part 1 point out that the construction of a self in rhetoric is dangerous stuff. The issue is not the audience, but the self being represented. Behn's letters suggest that the self represented in and through letters is itself a fabrication; as Gerald MacLean puts it, "after all, identity is only ever a rhetorical strategy, available as a trope."[25] When taken together with the other two parts of the novel, Part 1 also suggests that the idea that there is a stable self to represent via letters poses a deeper problem. Behn's use of epistolary form to question the nature of the self and the representation of the self thus may have been rendered more obvious or accessible to her because of the epistolary amatory tradition, but epistolarity's problems when it came to the nature and the representation of the self transcend familiarity and formality, philosophy and sex.

This idea of performance through rhetoric is also crucial to understanding Behn's use of epistolary form in *Love-Letters*. Critics of epistolary forms, particularly the epistolary novel, have often taken the letter as an authentic representation of the inner self.[26] Likewise, there is a general acceptance of the thesis that women and epistolary forms were forced together during the late seventeenth and eighteenth centuries. As Thomas O. Beebee succinctly puts it, "During the eighteenth century, novel, letter, and letter-novel became feminized literary forms."[27] Amanda Gilroy and W. M. Verhoeven point out, however, that the "most historically powerful fiction of the letter has been that which figures it as the trope of authenticity and intimacy, which elides questions of linguistic, historical, and political mediation, and which construes the letter as feminine," a fiction built on "the persistence of a rhetoric that equates epistolary femininity and feminine epistolarity."[28] In fact, as Gilroy and Verhoeven observe, the representation of subjectivity in epistolary form is highly conventionalized, and the conventions that succeed best are the ones most effectively "commodified" and "fetishized," to borrow their terms. Furthermore, they note, representation in epistolary form is "historically and culturally specific: the letters that come to us, and the fictions/histories that we write about them, rely on and bear the traces of particular historical practices."[29] Behn's amatory or familiar letters also exist within a larger discourse of the letter.

The first part of *Love-Letters between a Nobleman and His Sister* recognizes letters as a performance of self, a performance undertaken by male and female correspondents using rhetoric. Behn uses the letters of Part 1 to show how a self is constructed through language, how it is made recognizable and comprehensible through the conventions of rhetorical representation. Philander has certainly already discovered this aspect of language for himself. As Donald R. Wehrs puts it, "Philander conveys the conventions of Restoration amorous discourse to create a *persona* capable of seducing Silvia, his sister-in-law."[30] Crucially, Philander is aware from the beginning that he is using "such Cant and stuff as this, which Lovers serve themselves with on occasion."[31] Having received a summons to the cabal, Philander writes to Silvia that he will not go: "No, let the busy unregarded Rout perish, the Cause fall or stand alone for me: give me but Love, Love and my *Silvia*; I ask no more of Heaven" he informs her, noting that "could you but imagine (O wondrous Miracle of Beauty!) how poor and little I esteem the valued trifles of the world" (62). Silvia in turn tells Philander that "while I Write methinks I'm talking to thee" (37).

Philander brilliantly uses the letters to manipulate Silvia. After attempting to persuade Silvia that "in my Creation I was form'd for Love, and destin'd for my *Silvia*, and she for her *Philander*," he asks alluringly, "shall we, can we disappoint our Fate, no my soft Charmer, our souls were toucht with the same shafts of Love before they had a being in our Bodies, and can we contradict Divine Decrees?" (16). In case she can resist this fable, Philander immediately follows it with something more threatening: "Or is't undoing, Dear, to bless *Philander* with what you must some time or other sacrifice to some hated loath'd object, (for *Silvia* can never love again) and are those Treasures for the dull conjugal Lover to rifle?" (16) Philander invokes the "cold Matrimonial imbrace," "the clumsey Husband's careless forc'd insipid duty's," and the "drudgery of life" (16). Later, he changes the meaning of the word "marriage," announcing that he is not married to Mertilla, because "my Soul was Married to yours in its first Creation; and only *Silvia* is the Wife of my sacred, my everlasting Vows," and he identifies Mertilla as "this fatal thing *call'd* Wife" (16, emphasis added). In a later letter, he again uses the combination of threat and enticement, this time reversing the order to begin with the menace and follow it with the lure of endearment. "You love me not," he asserts, and moves from almost playful name-calling—"false Charming Woman!"—to something more threatening—"I burst with resentment with injur'd Love"—to the near-renunciation of his passion: "you are either the most faithless of your Sex, or the most malicious and tormenting: Oh I am past tricks my *Silvia*, your little arts might do well in a beginning flame; but to a settled Fire that is arriv'd to the highest degree, it does but damp its fierceness, and instead of drawing me on, wou'd lessen my esteem." He artfully pulls back at the last minute, after frightening (or attempting to frighten) her with the loss of his ardor, concluding, "if any such deceit were capable to harbour in the Heart of *Silvia*, but she is all Divine, and I am mistaken in the meaning of what she says: Oh my adorable," he instructs, having attempted to soften her through fear, "think no more on that dull false thing a Wife" (27). Even that is not his real aim, simply a link in the chain of reasoning, because he really aims at "*Silvia*'s Company, alone, silent, and perhaps by Dark" (28). The letter concludes with a second iteration of the threat-withdrawal technique. Philander orders in a postscript, "*resolve to see me to night, or I shall come without order, and injure both.*" To underscore his capacity for violence, he explains that "*My dear Damn'd Wife is dispos'd of at a Ball* Caesario *makes to Night,*" using the phrase "dispos'd of" as a double entendre of violence. He ends the postscript with "*the opportunity will be luckey, not that I fear her jealousie, but the effects of it,*" portraying himself as a potential victim to diffuse the menace of the rest of it (27–28). Although Silvia seems unmoved by the threat of violence—she only "tremble[s] with the apprehension of what [he] ask[s]" because she is afraid to see him alone, at night, in her bedroom (29)—Philander's use of language to manipulate her exemplifies the danger of language and rhetoric deplored by natural philosophers seeking, in Thomas Sprat's words, a "close, natural way" of writing, "bringing all things as near the Mathematical plainness, as they can" that reveals and conveys, but never obscures or manipulates.[32] Philander constructs a self that is changeable, and he does so to manipulate what he knows is Silvia's changeable self.

There is some question whether Silvia is aware of the ways in which Philander is manipulating language for his ends, or whether she is using language in the same way. When she permits him to visit her alone, at night, in her room, then adds, "If I incline but in a Languishing look, if but a wish appear in my eyes, or I betray consent but in a Sigh; take not, oh take not the opportunity," does she really think he won't? Furthermore, when she finishes that sentence with "lest when you've done I grow raging mad, and discover all in the wild fit," she seems to be using the same tempt-and-threat technique that Philander uses in some of his letters (30). Wehrs and Maureen Duffy suggest that Silvia does not recognize what Philander was doing until she was betrayed, in Part 2.[33] In a way, however, it does not matter whether Silvia recognizes the emptiness of Philander's words; whether Silvia knows that her own conventionalized replies are equally empty of the meaning they are supposed to have; or whether Silvia is as active and knowledgeable a partner in this exchange as Philander—whether, one might say, she is aware that they are playing a game and by what rules. It is crucial, however, that in the exchange between Silvia and Philander, Behn makes it clear that at least one correspondent's letters are manipulating rhetorical convention and therefore manipulating rhetorical expectation, both Silvia's expectation (that if Philander writes like this, he loves her) and Philander's (that if he writes like this, he will get her into bed). It only takes one side of the correspondence to show that self-knowledge and the self's knowledge of others are both heavily mediated by systems of representation and interpretation, and those systems are readily manipulated.

The epistolarity in Part 1 thus posits the impossibility of escaping the unreliability of language's representation of the self. Since external manifestations like language are the only data humans have, people are doomed to make use of a manipulated, corrupt data set as we try to understand other people. As early as the 1660s, Todd notes, Behn recognized the "power of language to create and deceive."[34] Behn was still exploring this point at the end of her career; writing about *Oroonoko* and *The Widdow Ranter*, Margarete Rubik suggests that in these texts Behn shows that "Like all the other criteria, language fails to provide a stable basis of identity."[35] *Love-Letters* also exposes language's inability to represent fully and accurately. The self cannot be known, or not reliably known, because it is made evident through the conventions of representation that are controlled by the individual who writes the letter. Those conventions are also, Behn points out, at least to some degree shaped by the society in which the individual lives. Silvia and Philander both allude frequently to how the "virtuous woman" or "virtuous man" is constructed by social expectations. "I can't forget I'm Daughter to the great *Beralti*, and Sister to *Mertilla*, a yet unspotted Maid, fit to produce a race of Glorious Hero's!" Silvia writes, attempting to define herself (25). And ultimately, it is the society of readers and writers that generate the conventions for representing love in writing.

While language is too manipulable to be reliable, Behn posits and then demolishes the idea that the body can represent the self, that its "rhetoric" is accurate. Silvia repeatedly points out the difference between the self manifested through writing and the self manifested through the body. "The Rhetorick of Love," as she calls it, is "half-breath'd, interrupted words, languishing Eyes, flattering Speeches, broken Sighs, pressing the hand, and falling Tears," and it is "with these soft easie Arts, that *Silvia* first was won" (33). Even when she thinks that she is "talking to" Philander when she writes him, she admits that "still, methinks words do not enough express my Soul, to understand that right there requires looks; there is a Rhethorick in looks, in Sighs and silent touches that surpasses all!" and furthermore, there is a way of speaking words, rather than writing them, that "may express a tenderness which their own meaning does not bear" (37, 38). Behn repeatedly reminds the audience of what Silvia forgets, however: that the body's rhetoric is not dependable. Silvia recognizes that a man false to his prince can be false to his mistress (23, 39–40) and Philander admits that he uses public behavior to hide his love for Silvia (20). Philander describes how Mertilla's infatuation with Cesario was increasingly revealed in and by her behavior despite Mertilla's best efforts, although her behavior originally successfully obscured her emotions (18). Taken as a body of letters (pun intended), Part 1 reveals how all rhetoric of the self, textual or material, involves performance.[36]

So much for the rhetoric used to capture or represent the self. What then of the self? For Behn, whether deliberately self-constructed this way or not, the self is an unstable, unknowable entity. Philander has already wavered considerably before the first letter, having been so in love with Mertilla that he practically kidnaps her (Silvia calls it a rape [26]), ceased to love his wife, and come to love her sister. He is also politically unstable, as he is of the "cabal" and even a significant recruiter to it (26) and yet entirely disinterested when the business of rebellion and usurpation conflicts with the business of sex with his teenaged sister-in-law. Part 1 is generally more interested in Silvia's instability, however, since as the object of both seduction and suspense—will she? Won't she?—her consistency, and in what cause, are the motivation of the narrative. As Silvia describes herself, "[O]h you cannot think *Philander* with how much reason you call me fickle Maid, for cou'd you but imagine how I am tormentingly divided, how unresolv'd between violent Love, and cruel Honour: You would say 'twere impossible to fix me any where; or be the same thing for a moment together" (25–26). This letter goes on to describe in considerable detail and political language the unstable self:

> There is not a short hour past through the swift hand of time, since I was not despairing raging Love, jealous, fearful, and impatient; and now, now that your fond Letters have dispers'd those Damons, those tormenting Councellors, and given a little respit, a little tranquility to my Soul; like States luxurious grown with ease, it ungratefully rebells against the Soveraign power that made it great and happy; and now that Traytor Honour heads the mutiners within; Honour whom my

> late mighty fears had almost famisht and brought to nothing, warm'd and reviv'd by thy new protested flame, makes War against Almighty Love! and I, who but now nobly resolv'd for Love! by an inconstancy natural to my Sex, or rather my fears, am turn'd over to Honour's side: So the despairing man stands on the Rivers Bank, design'd to plunge into the rapid stream, till coward fear seizing his timerous soul, he views around once more the flow'ry Plains, and looks with wishing eyes back to the Groves, then sighing stops, and cry's I was too rash, forsakes the dangerous shore, and hasts away. (25)

In this passage, Silvia describes how she moves quickly and also fluidly among different moods, characterizing that change in emotion in military and political terms—counselors, rebels, a town under siege, and so forth—to link the changing self to a changing political situation and all that might be at stake. Similarly, after the unconsummated tryst, Silvia describes two disordered, changeable selves, herself and Philander, turning repeatedly to "Madness" and "frenzy" as her theme to characterize her oscillations between calm and wildness, composure and uncontrollable weeping and tearing her clothes (63–69). But Silvia also changes with time. Over the course of the correspondence in Part 1, after all, she shifts from some kind of sexual resistance to whole-hearted sexual coconspiracy. Even the self seeking consistency and stability, the self "nobly resolv'd" (25) cannot will itself to constancy and rationality. As Silvia demonstrates, even its self-knowledge is inconstant and inconsistent. In this case as well, when viewed *in toto*, the letters of Part 1 reveal that the self is profoundly inconsistent.

This inconsistency can appear even within the space of a line, suggesting the rapidity of change. "'[T]is in vain (my too charming Brother) to make me insensible of our Alliance," Silvia writes, shifting between protestations ("it is in vain") to near-capitulation ("too charming") and back to resistance ("Brother") (13). A letter from Silvia to Philander before their first and unconsummated night together begins with "what pains and Pantings my heart sustain'd at every thought that brought me of thy near approach," imagines her father surprising them in her bedroom and moves to what Mertilla will think. Her mention of "*Mertilla* my Sister, and *Philander*'s Wife?—Oh God! that cruel thought will put me into ravings" triggers a stream of complaints against Philander, including the wholly irrational "Ah *Philander*, could you not have stay'd ten short years longer?" for Silvia to become a teenager, then fantasizes about Mertilla and Philander in bed together, shifts to calling Philander "thou Charming object of my eternal wishes," and hastens to a furious ending with "Yes you lov'd [Mertilla], false as you are you did, perjur'd and faithless. Lov'd her;—Hell and confusion on the world; 'twas so—Oh *Philander* I am lost—" and the editor notes, "*This letter was found in pieces torn*" (49–51). The editor's report that Silvia has torn up the letter contrasts with the missive's opening, which is about stability and commitment. Silvia starts by explaining that she has Philander's "tablets" and their key in which to write to him (49), signifying a wholeness and boundedness of self belied by the letter's contents and then by the action of shredding the document that she had just promised to create and protect with a lock and key. Significantly, here Behn needs a narrator to authenticate the instability with the report that the letters were found torn up, otherwise the final integrity of the letter might suggest too much control on Silvia's part. In other words, Behn needs a third-party witness to corroborate that Silvia's self is unstable because if the letter survived her paroxysms of emotion, she might appear stable. Overall, almost any letter in Part 1 illustrates changeable emotion and therefore the unstable self, the self that because changeable and complex, is never entirely known to its author.

The intervention of a narrator at this point highlights the problems of epistolary form when it comes to the body. Although letter writing is a physical activity, it is an activity that uses language to represent the immaterial, such as emotion or the memory of an event. When Philander writes to Silvia about her father's putting his penis into Philander's hand (61), the event is not actually happening in the narrative; what is happening, is that Philander is writing the memory of a physical interaction. Silvia's tearing up of the letter is an irruption of physicality that reveals a crucial problem with epistolary form: it cannot represent the full body-mind, material-immaterial combination that comprises the self.[37] As Ros Ballaster and Alvin Snider point out, however, Behn was very interested in the interaction of the body and the mind, an aspect of Lucretian ideas she found particularly significant.[38] The body's materiality means that it behaves according to principles. It cannot, in other words, be disregarded as experimental philosophers disregard it, either in using themselves as experimental subjects or in claiming to be able to train themselves to ignore their desires in probing the secrets of nature. For Lucretius, the materiality of the body means that it is an inseparable part of nature. It also means that the body's workings are indeed relevant to understanding the self and more broadly, nature. For Behn, this means that sexuality, particularly desire, are important factors in understanding the self: "Sexuality precipitates a disintegration of the self that undercuts the binary opposition of reason/emotion, mind/body."[39] As

Snider shows, for example, Behn's poem "To Daphnis," ostensibly celebrating Creech, "provides a site for the representation of the body as vulnerable to inspiration and passion, as subject to internal and external forces that reduce the autonomy of individual agents."[40] This element of the self necessarily had a significant impact on the self and needed to be represented. While scholars have focused on Behn's use of Lucretian ideas about the body in poetry, here in *Love-Letters* the interjection of a narrator in Part 1 and Behn's eventual dependence on it to explore the nature of the mind-body combination that is the self, suggests that for Behn, epistolary form was inadequate for this representation of the body but narration was not, or was less so.

Taken on its own, then, Part 1 comments on who we are and how we manipulate ourselves, others, and the conventions for representing those beings. Epistolary form focuses on representation as an issue more than on the actuality of experience, because of the nature of the Restoration letter. But as the interjection of the editor's report in this episode suggests, there are things epistolary form cannot do that an outside perspective can. Behn's increasing use of a narrator in Parts 2 and 3 allows her to continue to investigate the unstable, incompletely knowable self, including its existence as a combination of mind and body, and the techniques for representing it, with all their attendant advantages and problems. The shift to narration also, however, allows Behn to consider the consequences to the individual and society that an unstable self, combined with the erroneous belief in a stable, unified self, might engender.

At perhaps the simplest level, the inclusion of narration allows Behn to shift the focus from the representation of the self to the disjuncture between that representation and what people are thinking and feeling as they represent themselves in that way. In other words, using a narrator allows Behn to focus on the gap between representation and self *qua* gap, and it allows Behn to focus attention on either side of that gap: on the representation and on the self. The narrator shows us what characters are really thinking as they are doing certain things, whether they are performing or interpreting that performance. When Octavio writes Silvia a letter with expensive bracelets enclosed despite her insisting that he never write her again, the narrator reports that "*Silvia*, notwithstanding the seeming severity of her Commands, was well enough pleas'd to be disobey'd" while Silvia's letter opens, "You but ill judge of my Wit, or Humour, *Octavio*, when you send me such a Present, and such a Billet, if you believe I either receive the one, or the other as you design'd" (160). Part 2 is particularly preoccupied with this kind of manipulation of self-representation to achieve selfish ends: Philander seduces Calista, who is Octavio's sister; Octavio and Silvia wrangle over Philander's letter; Silvia and Antonett deceive Brilljard and Octavio; and so on.

To a certain extent, much of the narrator's function is the same in each of the second two parts of *Love-Letters*. Parts two and three continue to show how all the different languages of representation, including behavior and dress, are unreliable, and even manipulable. Silvia initially retains her masculine dress when she and Philander arrive in Holland because "pleas'd with the Cavalier in her self: [she] beg'd she might live under that disguise" (126). When Octavio gives his uncle a lampoon on Sebastian's attraction to Silvia, Octavio blushes, and after "considering his Face awhile," Sebastian accuses him of authoring the lampoon himself: "*I Fancy, Sir, by your Physiognomy, that you yourself have a hand in this Libel:* At which *Octavio* blush'd, which he taking for guilt, flew out into terrible Anger against him" (290–91). Sebastian may wish to read Octavio correctly, but he also fears a particular meaning and his fears cause him to interpret the text a certain way. Desire, Behn shows, has a similar effect. Reading Philander's cool first letter to Silvia from Cologne, Octavio "no longer doubted (and the rather because he hop'd it)...that *Philander* found an abatement" of his passion for Silvia (140). The acknowledgment of the role of Octavio's wishes occurs in a parenthetical aside, suggesting that the narrator but not Octavio realizes the role of wishful thinking in his interpretation of the letter. Unaware of the power of their desires, even when those desires are hardly hidden to themselves, characters cannot recognize the effect of those desires on what they consider knowledge.

Because she occupies her own position in the narrative, Behn's narrator functions to expose the dynamic: to put the emphasis on the interaction between people that is catalyzed by the representation of a self and followed by the [in]comprehension of that representation. The narrator often recounts conversations in which words and the way in which they are spoken are meant to give an impression of an inner self very different from the self that is there. When she explains that while trying to extract Philander's letter from Octavio, which Silvia has correctly concluded confesses the end of his passion for her, Silvia says, "'*I demand to see the return you have from* Philander, *for possibly*—' said she, sweet'ning her Charming face into a Smile design'd, '*I shou'd not be displeas'd to find I might with more freedom receive your Addresses*'" (195). "Thus she insinuated with all her female Arts," the narrator points out, "and put on all her Charms of Looks and smiles, sweetned her mouth, soften'd her Voyce and

Eyes, assuming all the tenderness and little affectations her subtil Sex was capable of" (195). Even when characters themselves read through the performance, the narrator functions to show the dynamic as it takes place, and to emphasize the dynamic itself. When Silvia replies seemingly angrily to Octavio's letter and gift of bracelets, Octavio disregards what she writes in her reply and instead takes as his text the fact that she writes back: "Let her, (said he in opening [her letter]) vow she hates me: *Let her call me Traytor and unjust, so she take the pains to tell it this way;* for he knew well those that argue will yield, and only she that sends him back his own Letters without reading em can give despair" (162). The narration shows how Octavio reads the situation and the woman correctly, choosing between two texts, her letter and her behavior, to do so. The necessity of the choice and the consequences of choosing well (or ill) are just as much the point here, as the question of whether Octavio will get Silvia or Silvia will get Octavio, since the terms on which their affair might be consummated differ depending on whose schemes succeed through correct reading.

With the outward manifestation of the self so readily manipulated and so unreliable, correct interpretation of another's represented self is crucial and yet profoundly difficult. The struggle is always to read properly, from Silvia's efforts to make true meaning from Philander's letter in Part 2 (141–43) to the long chain of men duped by Silvia in Part 3, including Octavio, Sebastian, and Don Alonzo. In the case of Philander's deceiving letter, for example, narrated, detailed descriptions of how each reader interpreted his or her letter emphasize the importance of a correct interpretation. Octavio "found, in that which *Philander* had writ to him, an Aire of coldness altogether unusual with that passionate Lover, and infinitely short in point of tenderness to those he had formerly seen of his, and from what he had heard him speak," and he sends Philander's letter to Silvia with one of his own, pointing out Philander's coldness so "her pride...wou'd decline and lessen her Love, for his Rival" (140). On reading the letter, Silvia "soon perceiv'd" that Octavio was right, but here the narrator provides a longer scene of psychological development. Silvia "repeated it again and again, and still she found more cause of grief and anger; Love occasion'd the first, and Pride the last" (141). The narration tracks Silvia through a sequence of emotional reactions, "considering" Octavio as a potential lover by reviewing his character, appearance, attractions, and wealth; planning to "conquer" him; and then writing a letter to do just that (133–34). The letter that follows this narration, however, functions as evidence of what has already been stated, that is, that Silvia intends to "torment" Octavio to "satisfie her pride" before she seduces him (142). The narration makes it possible to emphasize the struggle to understand, to render visible the confusion, the effort to interpret, and the reaction based on the results of those efforts. In this instance, the narrator's description and explanation expand and highlight Silvia's processes and therefore the difference between Philander's self-representation and Silvia's understanding of what is under the representation, the self discoverable only through proper reading. Philander and Silvia are hardly alone. Calista's attendant, Dormina, tells Calista that she has imagined Philander: "*your head runs of a young man, because you are married to an old one; such an Idea as you fram'd in your wishes, possest your fancy, which was so strong (as indeed fancy will be sometimes,) that it perswaded you, 'twas a very fantom or Vision*" (234). Is it real or is it her perception? Behn uses the narrator to carry the question through both parts of *Love-Letters*. When Hermione presents Cesario with what she is told is an enchanted tooth-pick case, designed to secure his affections to her permanently, "From that time forward she found, or thought she found, a more impatient Fondness in him than she had seen before" (400). In such moments, the narrative acknowledges the power of desire to construct a perception, and the very real consequences of trusting such a "fancy."

If the representation of the self is unreliable, so too, according to Behn, is the self being represented. The narrator offers descriptions of inner conflict, particularly connected to desire and knowledge. From his first entrance Octavio is conflicted, feeling attracted to both Philander and Silvia in her boy's dress. The narrator reports that the "lovely person of *Philander*, the quality that appear'd in his face and mien oblig'd *Octavio* to become no less his admirer" and after meeting Silvia, "he felt a secret joy and pleasure play about his Soul he knew not why; And was almost angry that he felt such an emotion for a youth, tho the most lovely that he ever saw" (122–23). The narrator also explains the physical manifestations of emotional states, as when Octavio discovers to his joy and horror that his sister has become Philander's lover, thus securing Octavio a reason for revenge and a justification he can accept to reveal Philander's infidelity to Silvia. "The pressures of his Soul were too extream before, and the concern he had for *Silvia* had brought it to the highest tide of Grief; so that this addition [of Calista's infidelity] o're whelmed it quite," the narrator reports, adding that "he threw himself upon his Bed, and lay without sense or motion for a whole hour, confus'd with thought, and divided in his concern" (243). Here narration allows Behn to resolve the problem of the role of the body in constructing the self by permitting Behn to describe the physical manifestation of Octavio's emotional conflict at the time it happens without the implication of performance that attends the representation of the self in a letter. Octavio remains an instance of conflict and indeterminacy to the end, when he decides to take orders and consequently must surmount his feelings for Silvia and for Philander (376–77), just as Silvia, Philander, and Brilljard remain examples of instability and flexibility. In such cases, narration permits the recounting of events without the distraction of an audience, enabling the narrative to focus on the emotional state, even when that state is confusion or a state the character could not identify or understand, sometimes even as it develops.

The narrator's role in confirming ideas put forward in the epistolary section is underscored by the different use of letters in the same scenario in Parts 1 and 3. Part 1 provides the correspondence between Silvia and Philander when Silvia is locked in her room by her parents but when Silvia is detained by Octavio's uncle Sebastian, Part 3 only provides the narrator's summary of her correspondence with Octavio and the feelings it generates (291–92). This change in representation signals a change in emphasis, from the representation to the disjuncture between representation and self, from self-knowing to others' knowing about the self. The self in *Love-Letters* is hardly capable of true objectivity, detachment, or unity. Rather, it is turbulent, capable of irrational self-destruction, full of internal contradiction, and difficult if not impossible to know.

Equally significant to this structural shift is how a mix of epistolarity and narration allows Behn to expand the context in which knowing or what passes for knowing takes place. After Octavio interrupts Brilljard as the latter is molesting the unconscious Silvia, Brilljard relieves his "feelings" with a willing, not to say inviting, Antonett. "*Brilljard* departed," the narrator reports, but "how well pleas'd you may imagine, or with what gusto he left her to be with the lovely *Octavio,* whom he perceiv'd too well was a Lover in the disguise of a Friend" (149). Brilljard immediately encounters Antonett, "and as much in Love as *Brilljard* was he found *Antonett* an Antidote that dispell'd the grosser part of it, for she was in Love with our Amorous friend" (149). Antonett, however, "who persu'd her Lover out half jealous there might be some amarous intrigue between her Lady and him, which she sought in vain by all the feable Arts of her Countries Sex to get from him, while on the other side, he believing she might be of use in the farther discovery he desir'd to make between *Octavio* and *Silvia;* not only told her she her self was the Object of his wishes, but gave her a substantial proof on't" (150). Having concluded their negotiations, "After she had promis'd to betray all things to him she departed to her affairs, and he to giving his Lord an account of *Silvia*, as he desir'd" (150). The narration underscores the physical proximity of the characters to each other—Octavio walks in a door to meet Silvia and Brilljard, Brilljard walks out the same door to meet Antonett, wooing takes place on either side of that door, and so forth—but it also situates the encounter within a larger emotional context not only of each character but also of the network of individuals—Octavio, Silvia, Philander, Brilljard, and Antonett. The emphasis shifts from what someone knows to the network in which knowledge, and pseudo-knowledge, are created and in turn create effects. Behn's experiments with narration thus allow her to focus on the instability of the self and the question of what is knowledge, a change from the focus on the representation of that instability implicit in her earlier epistolary form.

The narrator in Part 2 allows Behn to show all these issues in the characters: they do not comprehend themselves, they do not understand each other, and they manipulate conventions for representation. In Part 3, two years later, Behn complicates the situation further by using the seemingly reliable narrator to explore how people deceive not only each other but also themselves. Trying to recapture Silvia's attention after she tells him that she knows about his affair with Calista, Philander "flatters, and she believes, because she has a mind to believe; and thus by degrees he softens the listening *Silvia*...and yet so well he dissembled, that he scarce knew himself that he did so" (344). Without the letters to give the sense that a character has an understanding of himself or herself, the only representation of that character's self-knowledge comes through the narrator, a difference in sources that allows Behn to highlight how poor that self-knowledge can be. In the same scene, Silvia "suffers all [Philander] ask's, gives herself up again to Love, and is a second time undone...tho', if she had at this very time been put upon her sober Choice, which she would have abandoned, it would have been *Philander*" rather than Octavio, whom she has just agreed to leave for Philander (344). In the vision that Fergusano stages for Cesario, the latter rapidly oscillates between seeking love or military glory (405–7), an ambivalence that reappears in the eponymous protagonist of *Oroonoko*, who twice is rendered literally motionless by the loss of his beloved Imoinda, once when the enemy army is attacking in Coromantien and once when he kills her in the forest (*Oroonoko*, 27–29, 61–62). In these cases, characters do not know themselves, or are unable to handle the person who they discover they really are. Furthermore, in such moments, this misprision has physical as well as mental consequences: Silvia has sex, Octavio is paralyzed on a bed, Cesario declares war. The mind matters, but the body matters, too.

Self-deception is also a problem in the narrator, whose perceptions are added to the mix. The narrator in Part 3 possesses the hallmarks of reliability and appears to believe herself to be reliable, but she is not. Even from her earliest appearances in the second part of *Love-Letters*, she makes all sorts of claims to authority, such as having seen things herself or having spoken with many reliable people for her information. She is witty, keenly observing, and proximate to but separate from events themselves. "[T]here are but very few who find it necessary to die of the Disease of Love," she notes tartly (259). Reporting that Silvia has tricked Brilljard and then threatened to kill him with a knife, the narrator says that Brilljard's "thoughts [were] on nothing but a wild confusion, of which he vow'd afterwards he cou'd give no account of" (158). The narrator's own experience serves to authenticate the narrative. She claims to have genuine documentation, informing the audience that Cesario "neither showed the Majesty of a Prince, nor Sense of a Gentleman; as I could make appear by exposing those Copies, which I leave to History" (436). When Sebastian's servants discover his murder, the narrator explains how the culture of the country accounts for the servants' failure to pursue Octavio and Silvia (305). She would know it, since she claims to have lived for some time in Flanders (380). What the narrator does not witness, she sometimes claims to have from someone who has, as when she says that "I have heard her Page say, from whom I have had a great part of the Truths of her Life," or when she repeats twice that she learned about Cesario's military prowess from "a gentleman" who served with him (388, 428). Thus she establishes her credentials and credibility. Or so it seems.

At the same time that the narrator claims reliable knowledge, however, she undermines those claims, a technique particularly in evidence in the third part. Small lapses, such as forgetting the name of a village (374), are really part of a larger unreliability, demonstrated by her biased evaluation of characters. Castigating Cesario after his final military defeat, she snorts, "he should have died like a *Roman*, and have scorn'd to have added to the Triumph of the Enemy" (434). Of the people who supported his initial invasion, she calls them the "dirty Croud" and "under World" (426). Cesario's Captain of the Guard is an "old Hereditary Rogue...a Fellow rough and daring" (430). Cesario's counselors are "Knaves and Fools" (438). Fergusano is a "Villain" (439). Even the briefest descriptions are evocations of the personality that makes them.

The narrator's unreliability is particularly significant when it comes to Octavio, to no small degree because it is well hidden, at least from the narrator herself. The narrator's failure to recognize her own biases about Octavio renders her history of him suspect. When Octavio sends Silvia a letter from Philander, he lies about his reasons for not carrying it to her himself and takes care to inform her that he recognizes Philander's cooler tone about Silvia, signaling both his awareness of it and ensuring that Silvia sees it. "I wonder the mask shou'd be put on now to me," Octavio writes in mock-naïve astonishment, adding disingenuously, "I hope you will make the same favourable constructions of it" that he does, and "not impute the lessen'd zeal wherewith he treats the charming *Silvia* to any possible change or coldness" although that is precisely what Octavio is encouraging her to do (140). When he discovers that his gullible, poorly educated, undefended sister is being seduced by Philander, Octavio level-headedly assesses her danger:

> he consider'd her young about eighteen, married to an Old ill favour'd jealous Husband…he knew tho she wanted no Wit she did Art, for being bred without the Conversation of Men she had not learnt the little cunnings of her Sex, he guest by his own Soul that hers was soft and apt for impression, he judg'd from her Confession to her Husband of the Vision, that she had a simple Innocence, that might betray a young Beauty under such Circumstances; to all this he consider'd the Charms of *Philander* unresistible, his unwearied industry in love, and concludes his Sister lost. (178)

Conveniently, Octavio immediately "concludes" there is nothing to be done for his sister although Philander has not completed the seduction, and rather than attempt to save his sister, Octavio uses the occasion to vilify her and Philander and therefore to justify his own efforts to gain Silvia's affections: he mentally urges Philander to "*rifle* Calista *of every Vertue Heaven and Nature gave her, so I may but revenge it on thy* Silvia!" and "Pleas'd with this joyful hope he traverses his Chamber" (178). The narrator goes on to savor the beauty generated by Octavio's pleasure at the thought of Calista's seduction. Immediately following his "joyful hope," she described him "glowing and blushing with new kindling fire, his heart that was all gay, defus'd a gladness, that exprest it self in every Feature of his lovely face, his eyes that were by nature languishing, shone now with an unusual Air of briskness, Smiles grac'd his mouth, and dimples drest his face, insensibly his busie fingers trick and dress, and set his hair, and without designing it, his feet are bearing him to *Silvia*" (178–79). The vampiric pleasure Octavio derives from his sister's adultery makes him beautiful, or so the narrator finds him. This reaction to Octavio's abandoning his sister for his own purposes is the narrator's, however, not Behn's. In other texts, Behn presents Calista's situation as a tragedy for the woman. In "The Unfortunate Happy Lady," the villain of the piece is a brother who encourages just such a fall for his sister in order to benefit from it. In *The Rover*, Don Pedro attempts to dispose of his sisters against their will in pursuit of his own gain. But as the passage describing Octavio's appearance while contemplating his sister's seduction or the passage describing him as he takes orders indicate, the narrator has an attraction to Octavio that she does not have to other characters. After a detailed description of his "rich" dress, she moves to his body rather than his wardrobe:

> His Hair, which was long and black, was that day in the finest order that could be imagined; but, for his Face and Eyes, I am not able to describe the Charms that adorn'd 'em; no Fancy, no Imagination can paint the Beauties there: He look'd indeed as if he were maid for Heaven; no Mortal ever had such Grace: He look'd, methought, as if the Gods of Love had met in Council to dress him up that day for everlasting Conquest; for to his usual Beauties he seem'd to have the Addition of a thousand more; he bore new Lustre in his Face and Eyes, Smiles on his Cheeks, and Dimples on his Lips.... (382)

"For my part," the narrator confesses, "I swear I was never so affected in my Life, with any thing, as I was at this Ceremony, nor ever found my Heart so oppressed with Tenderness" (383). The body matters, and not just to everyone else; unknown to herself, it matters to the narrator. Her perceptions and therefore her knowledge are shaped by her desires, and her position in relation to the other characters makes her version of events compelling but not wholly authoritative. As I discuss shortly, she resembles no one so much as the narrator of *Oroonoko*, in fact, a narrator whose understanding of events is colored by her ability to know about them and by her feelings for the players, particularly Oroonoko himself.

While the idea that the self is opaque even when carefully self-scrutinized makes for psychological drama, Behn goes one step further, revealing the social and political as well as personal consequences of misunderstanding the self. Silvia, like Philander, is powerfully desirable and desiring, a combination that destabilizes the social institutions of her world, beginning with her family and expanding outward to the family, to gender, and to class. Other characters' failure to understand themselves, from poor ignorant Calista to the seemingly more sophisticated Octavio, from the inconstant Philander to the formerly static Sebastian, has a similar effect on family, marriage, primogeniture, and class. Rubik points out that such a maneuver characterizes *Oroonoko* and *The Widdow Ranter*, and *Love-Letters* offers yet another example.[41]

Behn played with the political consequences of unstable self-representation and unreliable self-knowledge in the first two parts, more strongly in the second than in the first as she forced Philander to flee Holland because of his political activities and had Octavio arrested as part of the knotty problems around identity and association. She recognized the political consequences of personal instability in Part 1 as related but secondary to the seduction narrative, such as when Silvia characterizes her self's instability in political and military terms. "Like States luxurious grown with ease, it ungratefully rebells against the Soveraign power that made it great and happy; and now that Traytor honour heads the mutineers within," she writes Philander, describing "Honour, whom my late mighty fears had almost famisht and brought to nothing, warm'd and reviv'd by thy new-protested flame, makes War against Almighty Love!" (25). Later, Silvia claims the right to administer to Philander a royalist lecture, concluding with "for you cannot intend Love and Ambition, *Silvia* and *Caesario* at once" (39–43). He certainly does, of course, and Silvia's notion that "the design is inconsistent with Love" proves false (38), but it is the personal rather than political intrigues that provide the focus of Part 1.

In Part 3, by bringing the narrator's perceptions into question Behn also brings the narrator's knowledge and way of knowing into question. No story, no data set, is reliable. That maneuver allows Behn to make the political consequences of this epistemology the primary focus of the narrative, bringing together the Whiggism, these notions of the self, and the problems they created. The sexual scandal that catalyzed the first part of *Love-Letters* was certainly political from the start. Lord Grey had been embroiled with the Duke of Monmouth from well before he eloped with Lady Henrietta Berkeley in the summer of 1682, and his political machinations continued through the winter of 1682–83 until he was imprisoned in the Tower of London and then fled to Holland in the summer of 1683 to avoid a gruesome execution for treason.[42] Lord Grey also was involved in the Monmouth Rebellion of 1685, although Monmouth's debacle hardly concluded the political problems of the 1680s. The decade was characterized by an intensifying sociopolitical crisis over the Stuart succession and monarchy, culminating in the "bloodless" or "glorious" revolution of 1688. The political philosophy evolving during the 1680s to grapple with this crisis shared key concepts and assumptions with much of the natural philosophy driving the members of the Royal Society.[43] Many of the political philosophers of the late seventeenth century were also natural philosophers—Thomas Hobbes as well as John Locke, who was also a Fellow of the Royal Society—and the "political version of selfhood" that Rachel Carnell tracks is closely related to, if not one aspect of, the larger self posited by natural philosophers.[44] Certainly philosophers as early as Bacon recognized the political significance of this idea of the self, and the epistemology accompanying it.[45] Thus, although Locke would not begin to publish his seminal works on the individual and on government until 1689, these ideas were hotly debated by different thinkers through the 1680s, particularly in response to the Exclusion Crisis (1679–81).[46] These political developments directly interacted with the emergence of the novel. As Ballaster and Carnell demonstrate, form and structure in early narratives such as *Love-Letters* reflect a concern with the political subject. Carnell argues that "many narrative techniques now associated with narrative realism were part of the cultural discourses competing to determine which political version of selfhood would be perceived as normative."[47] The political subject, in other words, is an aspect of rather than a separate entity from the idea of the self emerging out of natural philosophy at the same period, a self offering an irresistible challenge to the authors of what would become the early novel.

Small wonder, then, that Behn should recognize the implications of natural philosophy's idea of self for knowledge and the shape of the social and political order, and particularly over the course of the 1680s as those ideas continued to shape significant political events. Her analysis of Cesario's rebellion through the illusion of the stable self is a consideration of Monmouth's rebellion, and of the role of gender and the body in forming the self. It is also, however, an investigation of the larger issues of statehood, nationhood, and citizenship, as well as of the individual and the body politic. Albert Rivero, for example, notes that Behn's narrator in Part 3 of *Love-Letters* points out the constructedness of history itself, with political implications.[48]

For Cesario, of course, the question of deception and self-deception are crucial. There is the deception of the rebellion, first of all: it cannot take place without secrecy and deceiving the political order. In Part 1, Philander was the political seducer; in Part 3, it is Cesario. Brilljard reports that "his Person gain'd him more numbers to his side, than his Cause or Quality; for he understood all the useful Arts of Popularity, the gracious smile and bow, and all those cheap Favours that so gain upon Hearts" (397). Cesario buys on the cheap, for he gains them "without the expence of any thing but Ceremony," with "Affability," and with "bowing to the Crowd" (397). He pretends to modesty and to being interested in the welfare of others, but he really uses patronage as "Tools" and "he stops at nothing that leads to his Ambition" (397–98).

But while Cesario might manipulate his own appearance to seduce others into believing that he is interested in the welfare of France, he is at core unstable. Sometimes he wishes to seize control of the country, other times he wishes to spend all his time with Hermione. Nor he does understand himself. Although "it was not long, since it was absolutely believed by all, that he had been resolved to give himself wholly up to her Arms; to have sought no other Glory than to have retired to a Corner of the World with her, and changed all his Crowns of Laurel for those of Roses: But some stirring Spirits have roused him anew, and awakened Ambition in him, and they are on great Designs, which possibly ere long, may make all *France* to tremble" (400–401). The description of Cesario itself vacillates and is itself hard to pin down. Brilljard, explaining the situation to Silvia (and Behn's readers), moves smoothly between describing Cesario as devoted and besotted and describing Cesario as plotting an armed coup. The colon between "roses" and "but" is a transition, a punctuational gateway between one sense of Cesario—in fact, one self of Cesario—and another. Although Cesario appears to be "rul'd and govern'd" by Hermione "as she pleased," at the same time, "He has perpetual Correspondence with the Party in *Paris*, and Advice of all things that pass," and he is planning an invasion of France (400–401). Certainly there are signs that Cesario's self is inconstant of purpose: as Brilljard tells Silvia, the conspirators "feared nothing but the Prince's Relapsing, who, now, most certainly preferred Love to Glory" (402). Whether he is enchanted by Hermione's magic toothpick case or simply besotted with a woman who believes in magic, Cesario is not particularly strong or critically minded (400). He does not wish to "take on him" the "title of King" but "those about him, insinuated into him" that he should do it and "So that believing it would give Nerves to the Cause, he unhappily took upon him that which ruined him" (426–27). Having promised other supporters that "his design was Liberty only, and that his end was the publick good," he turns out to be seeking the crown for himself after all (427). Nor does the narrator stop there, but goes on to consider the different ways in which Cesario's decision alienated crucial factions of his power base, thus leaving him militarily as well as politically weak. When highly focused, localized action is called for, Cesario can do it. After beginning his rebellion, "to say truth," the narrator reports, "*Cesario* in this Expedition show'd much more of a Souldier than the Politician: His Skill was great, his Conduct good, expert in Advantages, and indefatigable in Toils" (428).[49] But when larger claims are made upon him—political vision, long-term fortitude, and so on—there is little there. He does not simply present a misleading self to others. He also does not possess a stable self to hide under a deception.

Behn complicates the question of Cesario's self by infusing the situation with more layers of indeterminacy. Cesario may think that he is in charge of his own rebellion, but he is being controlled by his mistress, Hermione, who "kept her Illustrious Lover, as perfectly her Slave as if she had ingag'd him by all those types that Fetter the most circumspect, and totally subdued him to her Will" (397). Hermione, however, is also manipulated, although not by Cesario but by Fergusano, the Scotsman who is from Germany, the man who "was once in Holy Orders" and is now a "Sorcerer," and a rather indeterminate person in his own right (399, 398). So Cesario controls his followers, but Hermione controls Cesario, and Fergusano manipulates Hermione and through her, Cesario, although Fergusano, "most wonderfully charmed with the Wit and masculine Spirit of *Hermione*," chooses to subordinate his actions to her desires (399). Just who is in charge here? The person in whom rebels place their faith is neither stable nor in command of himself; the person ostensibly in charge of him is neither in full command of Cesario nor of herself; and the person evidently with power over her is choosing to let her have power over him. There is no center to the cabal, no final resting point for authority, no reality for followers to accept. This illusion of a stable, reliable self thus has serious consequences for Cesario and also for the conspirators and for awhile, the country, which is plunged into civil strife. People lose fortunes, are exiled, and die as a result of a misplaced faith in Cesario.

The progression from epistolarity to full narration in *Love-Letters* thus reveals Behn's ongoing, intensifying concern with the notions of the self and the epistemology built upon it that was increasingly central to social and political developments, institutions, and systems in the 1680s.[50] Aware of the political consequences of the definition of the subject presented by philosophers of the day, in *Love-Letters* Behn tests and finds dangerously wanting not only the ideas of the self but also the epistemology increasingly based on them.

But *Love-Letters'* structural innovation comprises only one aspect of Behn's investigation of the nature of self and the role of representative modes in constructing that self and others' understanding of it. Like the personae employed by philosophers such as William Harvey and Robert Hooke or the voices of philosophical epistles in the *Philosophical Transactions*, the narrators of Behn's shorter fiction are present by being involved emotionally in their stories. At the end of "The Unfortunate Happy Lady," when Philadelphia's villainous brother has finally been thrown into debtor's prison, the narrator remarks that "finding no Money coming, nor having a prospect of any, the Mareschal and his instruments turn'd him to the Commonside; where he learnt the Art of Peg-making; a mystery to which he had been a stranger all his life long, till then."[51] In "The Adventure of the Black Lady," the narrator introduces the "Vermin of the Parish, (I mean, the Overseers of the poor, who eat the Bread from 'em," and plays on their proper name, the "Overseers of the Poor (justly so call'd from their over-looking 'em)."[52] Behn's narrators also provide moral evaluation, opposing "white Devils" to "the fair Innocent" in "The Unfortunate Happy Lady" (370) or pointing out that the unworldly but honorable Black Lady is a "Fair Innocent (I must not say Foolish) one" (316). Behn's narrators know their material and take it seriously, performing expertise and investment to banish doubt about what they recount, but they are not reliable. The narrator of "The Unfortunate Happy Lady" opens by announcing, "I cannot omit giving the world an account, of the uncommon villany of a Gentleman of a good Family in *England* practic'd upon his Sister, which was attested to me by one who liv'd in the Family, and from whom I had the whole Truth of the Story" (365). This announcement seems to be designed to establish the reliability of the narrator's account, but it also is a report of a report, "one who liv'd in the Family"; in an environment where authority and eyewitnessing is so important, such an equivocal position is worth noting. Similarly, the narrator of "The Adventure of the Black Lady" introduces her story with "About the beginning of last *June* (as near as I can remember)" (315). Here, again, Behn draws on the idea of the authority of the eyewitness only to undercut it with a dubious memory. What else might this narrator not quite remember? Such narrators, however they claim the authority of intimate knowledge, also serve to point out the ways in which even an eyewitness can be unreliable.

Behn's fiction also uses techniques generated by experimental philosophy's notions of self to question the self: what it is, what it means to be "inside" or "outside" oneself or events or a culture, and the impact of experience on interpretation—of others, of events, of one-self. Toward this end such works, like *Love-Letters*, often use the autobiographical interjection to destabilize rather than to establish firm authority. In *The History of the Nun* (1689), for example, the narrator interrupts the exposition with details of her own life. "I once was design'd an humble Votary in the House of Devotion," she reports, "but fancying my self not endu'd with an obstinacy of Mind, great enough to secure me from the Efforts and Vanities of the World, I rather chose to deny my self the Content I could not certainly prom-ise my self, than to languish (as I have seen some do) in a certain Affliction." But the narrator then questions that decision: "tho' pos-sibly, since, I have sufficiently bewailed that mistaken and inconsider-ate Approbation and Preference of the false ungrateful World, (full of nothing but Nonsense, Noise, false Notions, and Contradiction) before the Innocence and Quiet of a Cloyster."[53] The story continues from there, uninterrupted by further personal testimony.[54]

What is the function of such interjections? In the case of the aside in *The History of the Nun*, the content of this interjection works the-matically with the primary narrative, just as the autobiographical interjections in *Love-Letters* also support larger issues of identity and reliability explored by the narrative. Isabella, the young woman who joins a cloister at thirteen without realizing that she is capable of great passion, eventually flees the convent, marries, thinks she is widowed, remarries, and murders both first and second husband to retain the best that the world has to offer: wealth and the "fame" accorded an unblemished reputation. Like the narrator, Isabella makes a decision that she regrets later, although it is the opposite one. That the narra-tor and Isabella make opposite decisions that turn out to be the wrong one, or might be the wrong one, suggests that it is not the decision but the deciding that is at issue. Neither Isabella nor the narrator fully appreciated either her own nature or the nature, especially the ben-efits, of either a cloistered or a worldly existence; consequently, they both come to rethink their interpretation of the data and conclusion.

In the primary narrative, furthermore, Behn emphasizes the impor-tance of the position from which Isabella made her decision. Although Isabella's father and his sister, the abbess, supposedly offer her every opportunity to decide her future for herself, the position from which she judges—inside the cloister, with nunnery life the norm since she was a toddler—means that the comparison is uneven, between the familiar and the unfamiliar. Hence, while her father ostensibly tried to make the choice fair, by normalizing one empirical reality, he made the other alien. Similarly, although the aunt was charged with

offering Isabella an uninfluenced choice, and although the aunt genu-
inely thought she was doing so, in reality she provided information
that would encourage Isabella to choose the convent. As the narrator
observes, "the *Abbess* was very well pleased, to find her (purposely
weak) Propositions [not to become a nun] so well overthrown" (215).
Anxious as they are to secure a position on the margin for Isabella to
observe both sides—she is in a convent but often goes out, for example—her father and aunt cannot do it. By pairing Isabella's choice
with the narrator's, who decided in favor of "the world" from within
it, Behn challenges the ideas that the self can achieve detachment from
the circumstances being observed, and that observation can be uninfluenced by desire (in the case of the father and the aunt) or familiarity (in Isabella's case).

In *Oroonoko* these explorations of the self—what kinds of knowledge it is capable of, the nature of its relationship with the body,
what its authority is and ought to be—becomes most nuanced, trou-
bling, and complex. Emphasizing the problems of authority and the
individual, *Oroonoko* begins with and focuses on truth claims centered on the person of the narrator. "I do not pretend," the narrator
announces in her first sentence (8). She calls her work a "History,"
insists that Oroonoko is not a "feign'd Hero," and says her story was
not "manage[d]" by "Fancy" (8). Nor, she continues, "in relating the
Truth," does she "design to adorn it with many Accidents, but such
as arriv'd in earnest to him," a story with "enough of Reality to support it, and to render it diverting, without the addition of Invention"
(8). One sentence, in short, crammed with claims to truth—"history,"
"truth," "reality," "earnest"—and the rejection of falsehood—
"feign'd," "fancy," "invention"—immediately followed with the first
sentence of the next paragraph—"I was my self an Eye-Witness to a
great part, of what you will find here set down; and what I coul'd not
be Witness of, I receiv'd from the Mouth of the chief Actor in this
History, the Hero himself" (8)—with its claims to veracity. The story
is true because it is recounted by a reliable source, in other words.
Here, then, Behn hangs the authority, the factuality of the narrative
on the person of the narrator and the markers of authenticity: here is
what she saw and what she brought back for her audience to see in
addition to her narrative.

But as Margarete Rubik points out, identity in *Oroonoko* is
characterized by instability and multiplicity, whether that identity
is of a group such as the Caribs or the English or of an individual,
such as the narrator.[55] As in the short fiction and key moments in
Love-Letters, such autobiographical passages raise issues within the

primary narrative and in conjunction with it. The claim that "I was my self an Eye-Witness," used to certify the accuracy and authenticity of the events about to be recounted, has provoked numerous attempts to prove or disprove the veracity of this claim vis à vis Behn's own life (8). It seems safe to say at this point that the historical Aphra Behn did do some of what the narrator claims, but it is very unlikely that she did them all. That, however, is also the case for the narrator. The narrator's claim that "I was my self an Eye-Witness to a great part, of what you will find here set down" is thus only partly true both historically—for the author—and narratively—for the narrator—as neither author nor narrator was present for everything "set down." From the beginning, Behn introduces the possibility that the "I" of recorded experience and the "I" that had the experience are not the same. There is a gap, that is, between the "modest witness" who did the witnessing and the "modest witness" represented as having done the witnessing, the same gap that Behn increasingly exposes and investigates with her growing use of a narrator in *Love-Letters*.

This disjunction between narrator and author, narrator's experience and author's experience, is augmented by the materiality of the narrator. Unlike the essentially disembodied narrators of Behn's shorter fiction, *Oroonoko*'s narrator possesses a history, a family, a body, a wardrobe, and so on. Philosophers such as Robert Hooke, William Harvey, Robert Boyle, or Antonie van Leeuwenhoek used the sense of their material existence to command authority in their writing. Following this technique, the gift of a body in *Oroonoko* ought to enhance the narrator's tale's reliability, its accurate factuality. In *Love-Letters*, Behn used libertinism to investigate the role of the body in constructing knowledge. In *Oroonoko*, Behn uses other aspects of the body. Here Behn renders the narrator's physical self a burden; her materiality threatens her authority. Her weak constitution—"being my self but Sickly, and very apt to fall into Fits of dangerous Illness upon any extraordinary Melancholy"—directly affects her ability to know, as she leaves Oroonoko at a crucial moment because she is too ill to remain with him (64). Similarly, her female body also causes her to "fly down the River, to be secur'd" with the other English women (57).

Behn extends this idea to the narrator's experiences as well. Her physical absences at crucial moments challenge her claims to influence and importance and therefore her credibility. She says that Oroonoko promises not to hurt her—"and as for my self, and those upon that *Plantation* where he was, he wou'd sooner forfeit his eternal Liberty, and Life it self, than lift his Hand against his greatest Enemy on that

Place"—but she flees when his revolt begins (42, 57). At the end she says she could have saved him from whipping and execution, but she is conveniently absent when action is called for on these occasions, since "taking Boat, [she] went with other Company to Colonel *Martin*'s, about three Days Journy down the River" (64). As a result, she must recount the report of others who are not Oroonoko, although this is a source unacknowledged in her claims to authority at the start of the narrative.

Behn's point that the physical reality of the body necessarily affects the ability to perceive and interpret data is certainly at odds with the rhetorical posturing of Harvey, Boyle, or Hooke in their accounts of their work, and she challenges their notion that empirical experience in the form of encounters between one's own material existence and that of others contributes to authority.[56] But writers such as Harvey, Boyle, and Hooke who are using those strategies represent only one strand in natural philosophers' efforts to represent the self. Other authors such as Newton in the *Principia* attempted to disembody the philosopher's self from his body, avoiding the notion that the body was a flawed vessel or tool by avoiding the sense of a body in their writing. Rupert Hall suggests that Newton used math as well as rhetoric to make his findings seem inevitable, a way of compensating for the potentially corrosive influence of the body of the experimenter.[57] In the *Principia*, Newton deemphasizes the body of the experimenter to emphasize the objectivity and factuality of the data and conclusions. There is a first-person speaker, but he appears rarely if commandingly. Instead, Newton weakens the focus on a particular individual with other points of view, such as the first-person plural: "These quantities of forces, we may, for the sake of brevity, call by the names of motive, accelerative, and absolute forces." His primary rhetorical posture is a passive voice. Although Newton reiterates the three "quantities of forces" in a first-person plural, he introduces these quantities in a third-person declarative statement: "The quantity of any centripetal force may be considered as of three kinds: absolute, accelerative, and motive."[58] Such rhetoric depends heavily on the idea of the universal self, as it posits a sufficiently uniform audience that it would "consider" or do any other act exactly the same.[59] In embodying her unreliable narrator in *Oroonoko*, Behn takes on this other strand in natural philosophy: not the idea that the presence of the body in writing is authorizing (pun intended), but rather the idea that the *absence* of the body in writing is authorizing. Put together, Behn's narrators contend that the body does matter, that particularity therefore does matter, and that authority is challenged by the body.

The idea of "virtual witnesses," as Steven Shapin and Simon Shaffer call it, so central to the empirical method, also made bodilessness conceivable. Experimental philosophers may have been extremely careful in writing about the self in the text, but they also were careful to put the reader in a particular position where he became what Shapin and Shaffer call a "virtual witness" to their work. Newton's "we" represents one strategy, as his pronoun makes his reader a witness to the experiment and the reasoning that it provokes. In *Micrographia*, Hooke's magnificent images put his audience at the scope-end of the microscope, and his meticulous descriptions, as devoid of speculation as possible, recreate the experience of mentally recording while observing.[60] Shapin and Shaffer extensively document Boyle's rhetorical and visual techniques for establishing himself as a certain kind of authority so his readers would become a certain kind of audience: witnesses to his experiments.[61]

Behn's works, however, reject the notion that a readership is the same thing as a viewership. As a playwright and poet as well as a prose writer, Behn certainly knew how different genres worked, particularly how a performance-based medium like drama differed from a text-based medium like prose narrative. Like episodes in *Love-Letters* such as the scene of Octavio's taking orders, episodes in *Oroonoko* are profoundly sensory. Behn's opening paragraphs of *Oroonoko* are richly visual, such as the extended, detailed description of the Caribs or the vivid descriptions of fauna (8–9). *Oroonoko*'s descriptions also self-consciously appeal to the visual memory of her readers through allusions to England and the London theater, including the famous "Dress of the *Indian Queen*" (9). The significance of vision appears throughout the narrative, including its failure. The "misplaced" interjection about Imoinda's scars—"I had forgot to tell you, that those who are Nobly born of that Country, are so delicately Cut and Rac'd all over the fore-part of the Trunk of their Bodies" (40)—emphasizes who can see (the narrator), who cannot (the audience), and the dependence of the audience on the narrator for these details. The same holds for the descriptions of the natural world in Surinam, including her note that "in a Word, I must say thus much of it, That certainly had his late Majesty, of sacred memory, but seen and known what a vast and charming World he had been Master of in that continent, he would never have parted so Easily with it to the *Dutch*," or of Africans, as when she says, "And I have observ'd, 'tis a very great Error in those who laugh when one says, A Negro *can change colour*; for I have seen 'em as frequently blush, and look pale, and that as visibly as ever I saw in the most beautiful *White*" (43, 19). In moments

like these, the narrative contends that there is no replacement for seeing something firsthand. Similarly, the comparatively unvisual ending
contrasts with the abundant description of the opening, underscoring
that the narrator has not actually seen Oroonoko's execution.[62] While
the tale opens exhorting the reader to take the part of the narrator in
lamenting Oroonoko, by the end the narrative has also exposed the
reader's difficulty in joining the narrator. The reader and the narrator's experience cannot be the same, Behn's story insists, although her
narrator seeks to render them congruent.

The singularity of viewing and therefore of experience becomes
a singularity of understanding. The opening passage both asserts
absolute fidelity to the tale as it was presented or really happened,
and it acknowledges editing. The narrator notes that although she
is reproducing the story entire and faithfully, she also is "omit[ting],
for Brevity's sake, a thousand little Accidents of his Life, which...yet
might prove tedious and heavy to my Reader" (8). Behn emphasizes
the narrative as the product of this narrator. Her omissions and commentary also imply that her view is only one view—other characters,
such as Imoinda, Trefry, or Byam, would have a very different interpretation and also, crucially, very different access to facts, as they
were observers from a different position. Behn raises this point when
the narrator notes how little Oroonoko liked her religious teach-
ing but how much Imoinda appreciated it (41). Like the narrator of
Love-Letters, who is partial to Octavio, *Oroonoko*'s narrator also
has a partiality for a beautiful male, in this case Oroonoko, calling
him "great Man" from the start and a "mangl'd King" at the end (8,
65). In this text as well, Behn uses a narrator reveal how desire is
ineluctably influential. In fact, despite the claims of *Oroonoko*'s narrator to absolute reliability because she is an eyewitness, she declares
that she writes with a particular agenda: to justify and immortal-
ize Oroonoko (8). Nor can she reliably process data. She has ample
evidence that Oroonoko is a man of his word, but she cannot accept
that. When "he assur'd me, that whatsoever Resolutions he shou'd
take, he wou'd Act nothing upon the White-People"(41–42), he makes
a promise he keeps, as his rebellion intends escape, not revenge. But
the narrator, who has seen and heard him make this promise, has had
additional evidence of his veracity and honor, and is telling the story
to demonstrate his superior veracity and honor to the Europeans', still
flees when he launches his revolt. Why does the narrator not trust
Oroonoko? Perhaps she, like the captain who kidnapped Oroonoko,
cannot avoid believing that people lie, no matter what their ethical
framework.[63] *Oroonoko* suggests that the textual representation of

events requires interpretation and that therefore the text itself cannot be objective, whatever the mindset of the individual during the observation. Of course, *Oroonoko*'s narrator is no more cohesive than she is objective. While the narrative offers a sense of her as a material person with physical details such as her dress, her family, her constitution, and even the cut of her hair and her hat, the narrator does not give the reader an equally definitive sense of her own character. In the long paragraph where the narrator explains the rising tension between Oroonoko and the white male authorities, the narrator also replicates these tensions between herself and Oroonoko. "They" kept promising to let him go until Oroonoko distrusts "them," so "I was oblig'd, by some Persons, who feared a Mutiny" to speak on their behalf, which the narrator calls "giv[ing] him all the Satisfaction I possibly cou'd" (41). The pronoun confusion reveals the narrator's own confusion about her identity, and suggests either multiple and conflicting identifications or a changeable self. She identifies with women, with Christians, and with the classically educated (41), but she also distinguishes herself from the Englishwomen by bragging about how Oroonoko calls her his "*Great Mistress*" and from the African women by blaming them during the revolt (41, 55), criticizes Christianity (32–33), and separates herself from the other English colonists in Surinam (43). Rubik points out that these shifting identities destabilize the categories that impose identity: "The categories of race, class, and gender all prove illusory as the qualities attributed to any of these groups fluctuate disconcertingly," she notes.[64] The narrator's equivocal self may be a reflection of her social indeterminacy as the daughter of the late governor, but the narrator's multiple ways of defining herself also engages the ideas of the unified self.

This ambiguous self-positioning suggests that like Isabella, the narrator of *The History of the Nun*, or the narrator of *Love-Letters*, the narrator of *Oroonoko* does not perceive herself consistently or understand herself thoroughly. She is a paradox, wishing complete understanding of and centrality to events while at the same time trying to preserve partiality and marginality. As her opening indicates, she seems to believe that she has seen everything, that she has presented the unvarnished truth, and that her narrative presents facts that can stand on their own. Ultimately, according to *Oroonoko*, it is not possible to be a detached or consistent observer and it is delusion to think oneself so. The narrator is both in the middle of everything and on its margins as observer, both a part of primary groups (and there are several) and uncomfortably unlike them. She cannot be pinned down, and the ideas of reliability introduced by experimental philosophy require that without a set identity, a person cannot be acknowledged as an authority.

Part of Behn's art in *Oroonoko*, then, is its hall of mirrors, with the possibilities of an authentic self constantly receding. Furthermore, Behn points out the inevitable imaginative component to making sense of the data of the senses, or in narrative terms, to integrating experience into the self. Behn uses narrators to question any given self's ability to know reliably and fully. The self cannot occupy a position from which he or she truly knows; positionality is not only inevitable but also an influence, even when the thing to be known is the self. By representing narrators who claim to be (and in some cases think themselves to be) fully knowledgeable as incompletely knowledgeable and also incompletely aware of the limits of their knowing, Behn responds to the idea of the outsider in possession of reliable knowledge with doubt, if not outright skepticism. Such a person cannot really be what he or she thinks he or she is.

Taken together, Behn's narrators suggest an exploration or testing of the idea of a stable, reliable, objective self, if not a skepticism about it. Just as Behn recognizes the way that natural philosophy constructs its own key elements through language and spectacle, through pronouncement and performance, to achieve its own influence and ends, she also recognizes the constructedness of this self as well as its usefulness. Even if experimental science and with it, the stable, unified self is a grand performance in Behn's hands, the performative, spectacular element is unavoidable and crucial. As Jacqueline Pearson explains, "Behn has her female narrators humbly accede to the view that female creativity should be confined to certain fields, but this transparently ironic humility does not so much accept the conventional limitations as draw mocking attention to them. 'History' may be the locus of a specifically male authority, but male 'authors' are mocked by implication for the imperfectness of *their* authority."[65] While Pearson speaks of "history" here, a look at Behn's *oeuvre* shows that this formulation may be understood more broadly. For Behn, the claims of natural philosophy as the norm and the truth are problematic at best. They are also fascinating, however, offering ways to approach the issue of what it means to be an experiencing, thinking self and the question of how that self, and its tangled existence, is represented in writing.

Aphra Behn's narrative fiction reveals an author outside natural philosophy taking up the ideas of the "new science" and the rhetorical challenges generated by those ideas. The notions that a person can stand intellectually and emotionally beyond events, can therefore recount those events entirely reliably, and can use those events to present a credible explanation of the natural world (including humanity), appear to have raised compelling questions for authors of different kinds and different agendas. For fiction's writers, however intellectual or mercenary their goals, the entry of these ideas into the common consciousness, the popular discourse, offered significant opportunities for thought and text. Aphra Behn's narratives provide a valuable view of this process and a starting point for understanding how these ideas about the self helped shape the novel.

Chapter Three

The Fly's Eye: The Composite Self

As Anthony Wood laments in *An Instance of the Fingerpost*, Iain Pears's novel about the philosophical revolution in seventeenth-century Oxford, "[T]he infallible philosophical method seems inadequate when it comes to problems in which motion derives from people rather than dead matter."[1] Even in the twenty-first century, we are more accurate predicting the orbits of comets than predicting the motions of human individuals. Why the difficulty? If people are people just as apples are apples, why should the "infallible philosophical method seem inadequate" for explaining "motion deriving from people rather than dead matter"? And what about that word "seems"—is it ever possible to overcome that inadequacy and if so, how?

For Aphra Behn and writers with her preoccupations, the problem lay in the idea of the self. "Dead matter" has a certain integrity, stability, and predictability, but the self does not. Behn's narratives use point of view to expose both the instability of the self and the dangers of building on a false notion of selfhood. The problem of the unstable, unreliable self attracted other philosophers and other novelists as well, thinkers and writers with other questions and solutions. Among them was Jane Barker. Barker's narratives both recognize the impossibility of explaining, in Pears' terms, human motion and propose a solution for the limited ability to establish knowledge inhering to any individual self. Using frame narratives, Barker's narrators show that the self is capable of self-knowledge but not complete or fully accurate knowledge of others; for that, an individual must collect insight from a variety of unreliable sources to extract what is useful and synthesize it into something approximating—and only approximating—knowledge. Nesting one view of the world within another, Barker's Galesia trilogy accepts the fallibility of the self but proposes a remedy in the aggregation of fallible selves. Just as the fly's eye provides a view of the world, incomplete but the best the fly can do, so frame narrative offers a composite view of human nature that is both incomplete but the best that any single self can achieve.

Barker's narratives draw on a strain of thought in seventeenth- and early eighteenth-century natural philosophy that recognized the problems of the perceiving individual and addressed the need to remedy the flaws of perception and intellect bedeviling human beings. While there could be no doubt of a phenomenon—the bird in the air pump died, the comet appeared, the young man whispered sweet nothings, the young woman blushed—just what everyone perceived in the course of its happening was not inevitably exactly the same. Similarly, making knowledge from that data was an internal problem, a problem with and within the self, because the intellect was also fallible. For many English natural philosophers as well as their friends on the Continent, the solution to such problems was the group. Whatever the individual endeavor of the self might be, its discoveries could not be considered knowledge until corroborated by the community, until validated by a group of "modest witnesses." Knowledge of the event would therefore be comprised not of a single fact, but of an amalgamation of perceptions combined to make a·fact. For experimental philosophers, "solutions to the problem of knowledge are embedded within practical solutions to the problem of social order," Steven Shapin and Simon Shaffer point out in *Leviathan and the Air Pump*; knowledge has "consensual foundations."[2] Joseph Glanvill asks in *Scepsis Scientifica*, "For if we were yet arriv'd to certain and infallible Accounts in Nature, from whom might we more reasonably expect them then from a Number of Men, whom, their impartial Search, wary procedure, deep Sagacity, twisted Endeavours, ample Fortunes, and all other advantages, have rendered infinitely more likely to have succeeded in those Enquiries[?]"[3] The group, a combination of flawed minds and flawed perceiving organs, contributed to the making or recognizing of knowledge by dint of number, that is, of probability.[4] Even if the substance did not feel as cold to me as it did to you, we agree that it was cold. Add in George and Gracie's perceptions of cold, and now we have knowledge. Modest witness + modest witness + modest witness = knowledge.[5] As Donna Haraway points out, building an epistemology on the notion of a group of modest witnesses is as much a constitutive act as creating an epistemology built on the idea of a single "modest witness."[6]

Although scholars and critics have often focused on the "modest witness" as an individual, in fact the role of the individual was in tension with the role of the group. Peter Dear notes that part of the Royal Society's success lay in its aggregating force, with the effect that the "cooperative investigation of nature both shaped and was made possible by the new forms of natural knowledge."[7] Shapin and Shaffer's analysis of Robert Boyle's conflict with Thomas Hobbes exposes how much Boyle's epistemology depended on group consensus, and

Boylean experimentalism was also at odds with Newtonian mathematical individualism.[8] Consequently, over the period that Jane Barker was writing the Galesia trilogy—*Love Intrigues* (1713), *A Patch-Work Screen for the Ladies* (1723), and *The Lining of the Patch-Work Screen* (1726)—it was both possible to imagine an epistemology that emphasized the communal aspect of knowledge building rather than the individual experimenter or witness and to see how such an epistemology might be a form of resistance.

The problem of an individual's accurate perception was particularly pressing for women, whose correct understanding of others was crucial to social and often, literal survival in eighteenth-century England. The external data had its own integrity: the young man made certain statements of love. The self's processing of that data was less reliable: could the young woman tell if he were sincere? Toni Bowers points out that "anatomical vulnerability, social subordination, and violence" were not just a reality, but a "daily experience" for women. The perils of this position were immediate and pressing.[9] The effusion of amatory fiction during this period, exemplified by Eliza Haywood's inability to saturate the market with amatory narratives during the 1720s, suggests that there was an insatiable interest in narratives addressing such questions. Women writers certainly recognized the very real consequences for women if the generation of knowledge were considered the province of an individual, stable, unified self, particularly if that self did not actually exist. A woman was on her own when it came to interpreting others, and she could not trust herself to interpret accurately.[10]

While amatory fiction tended to explore these issues through theme and character, however, the early novel as an evolving genre also offered an opportunity to explore the problems of the individual, perceiving self through its structure. For Barker, framed narrative served to contrast the instability or unreliability of the self with the stability of the incoming data, and it allowed her to demonstrate the role of the group in delineating the known. Barker's Galesia novels employ a series of narratives, each one nested within another, to argue that the self was capable of self-knowledge but not capable of reliable, complete knowledge of anything or anyone beyond the self. Like Aphra Behn in chapter 2, Jane Barker uses narrative structure to reject the idea of the stable, unified self that is capable of creating full and reliable knowledge, but unlike Behn, Barker also uses narrative structure to suggest that insight can be created by combining the limited views of a collection of individual selves. In this framework, the most complete knowledge can be obtained only by synthesizing the perceptions and insights of a group. Scholars since Ian Watt, including those recovering women's writing, have placed the novel within an individualist tradition. Such approaches emphasize the novel as a representation of a singular—often exemplary, eventually "realist"— self, even when situating authors or novels in dialogic or sociable networks.[11] Although hardly the only instance of frame narratives used for this purpose, Barker's work poses a particularly acute challenge to this history of the novel and to the idea of the ready triumph of Lockean individualism. In rejecting the ideas of individualism that were gathering ideological and political force at the time, Barker advocates a collectivist epistemology for maximizing knowledge.

For biographical reasons, certainly, Jane Barker is a particularly apt example of a writer who understood the importance of accurate perceiving and the problems of the isolated self in constructing knowledge. Kathryn King observes that as an unmarried woman who was not only Catholic but Jacobite after 1688, Barker was a "member of four overlapping unprivileged minorities," populations increasingly disenfranchised by the changing social, political, religious, and economic systems of the late seventeenth and early eighteenth centuries.[12] Furthermore, Barker was a "learned lady," another isolating attribute, and spent some of her life in voluntary exile in France. At the same time, however, Barker managed to find larger questions and larger truths as she interrogated her experience and its meanings. "It is no small part of her achievement," King writes, "that she was able to create out of personal experience stories for a host of Others—Catholics, Jacobites, women, spinsters—who felt themselves excluded from official constructions of national identity in postrevolutionary Britain."[13] Following King's call to understand Barker and her work within the communities and networks she invoked and experienced, this chapter shows how novelists exemplified by Barker used narrative structure to investigate both the individual's very real flaws, and how the community, a collection of flawed selves, can compensate for those flaws to create knowledge.[14]

As Josephine Donovan has noted, by the seventeenth century women writers had been using framed narratives, what she calls the "framed-nouvelle" for some time. In the seventeenth and eighteenth centuries, its authors included Margaret Cavendish, Duchess of Newcastle; Delarivier Manley; and Sarah Fielding.[15] Donovan argues that this kind of structure permits the investigation of many selves, at least some of whom would otherwise have no opportunity to be presented, and combines those many selves into a larger, unified presence. But while Donovan tends to view the "framed-nouvelle" as the narrative structure of choice for women excluded from classical education and presumably the Scholastic tradition, its use by an author like Jane Barker, who did read Latin, suggests that its flexibility in presenting a variety of perspectives appealed to a range of authors who shared the interest in a "dialogic" narrative structure.[16] The framed novel generates an intra-textual dialogism among the frames, but it also permits the narrative itself to dialogue with its larger context.[17] In the case of the works discussed in this chapter, the larger context is the developing epistemology arising out of the philosophical revolution.

Conventionally, "framed narratives" have one or two frames within which other stories are told such as Mary Shelley's *Frankenstein*, Boccaccio's *The Decameron*, or Marguerite de Navarre's *Heptameron*. Often the frame works laterally: each of the framed narratives is contained within the larger frame, that is, each pilgrim in Geoffrey Chaucer's *Canterbury Tales* tells a story within the larger frame of the narrator's traveling with the pilgrims to Canterbury. In works like *The Canterbury Tales*, the framed narrative has a relationship with the framing narrative and often has a relationship with other framed narratives. The Miller's Tale responds to the Knight's Tale, which precedes it, and the Reeve's Tale in turn responds to the Miller's Tale, which precedes it, for example, but the Wife of Bath's Tale also can be read into the conversation despite being out of this sequence. Although *Love Intrigues* is fairly straightforward, with a frame narrative around Galesia's recounting of her life, Barker's narratives increasingly complicate the framing structure. By her final work, *The Lining of the Patch-Work Screen*, the number of framing narratives can be considerable, so instead of a story-within-a-story, Barker offers a story-within-a-story-within-a-story-within-a-story—and sometimes still more layers. All the frames work together to present an idea or explore an issue. Each framed narrative offers one perspective but only one; each is a piece of a larger puzzle that is more completely assembled when other framed narratives are used to illuminate the issue. Like the eye of the fly, brilliantly and strikingly illustrated by Robert Hooke in *Micrographia* (1665), each narrative provides one lens, one view, of an object or idea; a sense of that object or idea is only possible when the views through all the lenses are combined.[18] Put another way, where modest witness + modest witness + modest witness = knowledge, in Barker's work framed narrative + framed narrative + framed narrative = knowledge. Crucially, however, Barker's purpose is not to offer a particular item of knowledge or a specific insight. The multiplication of framing devices, and the transformation of a framed narrative into a framing narrative and back again, emphasizes the method by which knowledge is collected. This layering could be an effective strategy for works engaging with the epistemological developments of the early eighteenth century as it served to challenge the primacy of the unified, stable, reliable self and to offer a modification of natural philosophy's epistemology for addressing an alternative understanding of the self.

Discussions of point of view in the Galesia trilogy usually focus on Galesia and often consider ways in which Galesia does and does not possess a self like Barker's.[19] However, at the same time that Barker offers us a "recognizably modern individual who knows herself in large part through silent, solitary, even furtive acts of reading" who is the "interiorized, print-based subjectivity of an ordinary woman who finds herself propelled by her own 'uncouth' desires,"[20] Barker also offers us a narrator and other, subordinate characters with selves within narrative structures emphasizing the fact of different perspectives belonging to different people. Not all characters are equally profoundly drawn, and the different degrees of detail allow Barker to highlight how little people might understand or know about each other, and how important that limitation can be in understanding nature, human nature, and human society.

It was not necessary to be familiar with natural philosophy per se to be interested in the issues it raised, such as the nature of the self. Nevertheless, like Aphra Behn, Barker was familiar with the "new science" and she recognized significant issues in the philosophical revolution. Kathryn King and Jesslyn Medoff's extensive archival work offers several aspects of Barker's life that connected her to the revolution in natural philosophy. Barker's family seems to have been on the side of the "new learning." Barker's father worked with and for Robert Clayton and John Morris, whose innovative economic thinking revolutionized the banking and credit system in England. Through her mother's family, the Connocks, Jane Barker was related to Richard Lower, who was part of the group in Oxford during the 1660s that included Boyle, Wren, and Willis and who pioneered the study of the central and peripheral nervous system. Barker's older brother, Edward, trained in medicine at Oxford, a bastion of the new medical learning, and his appearances in Barker's writing indicate that they were close. Barker herself sold a cure for gout during the 1680s and claimed some success as a professional healer. In her biography of Barker, King offers a convincing case for Barker's expertise in empirical as well as academic medical practices.[21]

In addition to the biographical evidence, there is considerable textual evidence of her interest in natural philosophy. Most famous are Barker's "medical poems" in which, King suggests, Barker's treatment of Galenic and "scientific" approaches and conclusions to medicine "may represent a strategy for dramatizing the struggle to accommodate new scientific findings to older conceptual models belonging to the classical Galenic tradition" and "could be said to figure the confused, shifting, and contradictory play of scientific understandings during a time of paradigm shift."[22] Barker's poem, "A Farewell to Poetry, with a long Digression on Anatomy" from *Poetical Recreations* (1688) is a spectacular journey in verse through the body, notably the circulation system. Some of these poems were revised and reprinted in *A Patch-Work Screen for the Ladies* including "An Invitation to my Friends at Cambridge," which discusses conditions under which philosophy is best pursued, contrasting the frustration of unproductive isolation with the projected pleasure of productive community.[23] Barker explored similar epistemological problems in her fiction. Misty G. Anderson, for example, has argued that Barker's *Patch-Work Screen*, like Margaret Cavendish's *Convent of Pleasure*, is a reaction against Cartesian materialism and Lockean empiricism.[24] Certainly the allusions in her poetry and prose position her within the English experimental tradition.

Barker's interest in the "new science" runs throughout the Galesia trilogy. The bulk of Barker's overt endorsement appears in the first two narratives, *Love Intrigues* and *A Patch-Work Screen*. These books feature Galesia's brother, who is trained in the latest medical knowledge and techniques, and Galesia's own familiarity with the same body of knowledge, methodology, and ideology. Galesia frankly announces that this "Gentleman and Christian" (an echo, perhaps, of Boyle's *Christian Virtuoso* [1690]) has shared both his knowledge and his methods with her.[25] "[M]y brother continued to oblige my fancy," Galesia tells her friend, "and assisted me in *Anatomy* and *Simpling*, in which we took many a pleasing Walk, and gather'd many Patterns of different Plants, in order to make a large natural Herbal. I made such progress in *Anatomy*, as to understand *Harvey's* Circulation of the Blood, and *Lower's* Motion of the Heart."[26] Such is Galesia's dedication to this branch of learning that even after her brother dies, she peruses not just his books but also his notes despite the grief generated by the encounters with his handwriting and evidence of his thinking (*PWS*, 84).

Galesia has not just learned medicine from her brother, however. His more extensive legacy is the method of disciplined, regular study which Galesia adopts early in *Love Intrigues* and maintains throughout the trilogy. In *Love Intrigues*, Barker celebrates the study of nature as a rigorous, regular approach yielding practical, useful knowledge and defines nature in broad terms. Deprived of her favorite men—her brother and Bosvil—Galesia explains that "I return'd to my Studies; the Woods, Fields, and pastures, had the most of my Time, by which Means I became as perfect in rural Affairs as any *Arcadian* Shepherdess; insomuch, that my Father gave into my Power and Command all his Servants and Labourers."[27] Although she considers this responsibility "an Impediment to my Studies, yet it made Amends, it being itself a Study, and that a most useful one" (*LI*, 35). Barker here uses "study" and "studies" to refer to the medical knowledge shared by Galesia's brother as well as the practical farming knowledge that Galesia acquires on her own. She also uses those terms to refer to the method of gathering that knowledge both by reading and by steady, careful observation of the world outside of books. In effect, Galesia unites the two forms of witnessing offered by empirical philosophy, the witnessing of reading ("virtual witnessing," in Shapin and Shaffer's terms) and the witnessing of observation.[28] Her expansiveness also brings a number of ideas about knowledge and method together: Galesia's empirical knowledge of farming really belongs with her brother's academically based knowledge of medicine. In fact, in an aside about the importance of useful knowledge, Galesia sounds almost Baconian:

> The Rules to sow and reap in their Season; to know what Pasture is fit for Beeves, what for Sheep, what for Kine, with all their Branches, being a more useful Study than all the Grammar Rules, or Longitude and Latitude, Squaring the Circle, &c. The Former, according to the Utility of his Occupation, deserves to hold the first Rank amongst Mankind: That one may justly reflect with Veneration on those Times, when Kings and Princes thought it no Derogation to their Dignities. The Nobles, in ancient Times, did not leave their Country-Seats to become the Habitation of Jack-daws, and the Manufactory of Spiders, who, in Reproach to the Mistress, prepare Hangings, to supply those the Moth has devour'd, thro' her Negligence, or Absence. (*LI*, 35)

In an echo of the Royal Society's mission statement and Bacon's own emphasis on the useful,[29] Barker presents practical knowledge that affects the quality of life as far superior to the intellectual and purely academic knowledge of grammar or squaring the circle.

The idea that knowledge should be used to benefit others—exemplified by the scenario in which Galesia wishes to be alone but is paternally and ethically compelled to serve the family and its dependents—implicitly underpins Barker's larger idea that the contributions of others should be gathered to create knowledge. Both views are predicated on the idea that the self is not reliable: its abilities to perceive and to generate accurate knowledge are limited, regardless of the nature of the data being emitted by the natural world. Barker's novels are structured with these concerns in mind: her layering of narrators creates a patch-work of selves that, taken together, permit a complete and authoritative form of knowing. Thematically and structurally, Barker's novels explore the limits of what is knowable, repeatedly suggesting that the internal self is not as difficult to chart as external reality.

Barker's first novel, *Love Intrigues* has at its core the difference between knowing oneself and knowing what is beyond the self. *Love Intrigues* is presented as a narrative within a narrative: a frame narrator introduces Galesia, who then tells her own story in the first person to her friend Lucasia. Underscoring the difference in perspective is the primary conflict of Galesia's story, an unfortunate series of communications and miscommunications about their relationship between herself and Bosvil, the man she loves, that ends in his marrying another woman. Seen through Galesia's eyes, Bosvil's self is never really knowable. He says conflicting things about his feelings for Galesia and other women, and about his intentions toward these women, and he acts in conflicting ways. Early in their relationship, Galesia explains that "tho' he made no formal or direct Address to me, yet his Eyes darted Love, his Lips smil'd Love, his Heart sigh'd Love, his Tongue was the only Part silent in the Declaration of violent Passion; that between his cold Silence and his Sun-shine Looks, I was like the traveller [sic] in the Fable" (*LI*, 16). After he declares himself her lover and proposes marriage—"Many Days and Weeks pass'd, and several Visits he made with repeated Assurances of his Passion, still expecting the Coming of his Father" (26)—Bosvil goes to his father for the latter's blessing, but when he returns, "*Bosvil* came, but not my Lover: He came with greater Coldness and Indifferency than ever! No Ray of Love darted from his Eyes, no Sigh from his Heart, no Smile towards me, nothing but a dusky cold Indifferency, as if Love had never shin'd in his Hemisphere" (28). Shortly after he has proposed marriage but returned from an absence with very different behavior, Bosvil "puts forward" a friend of his as a husband for Galesia. Galesia then tests him with a letter, to force him into becoming comprehensible—either he must marry her or stay away—and he chooses the latter (30–34). Galesia often reflects on how mysterious he is and how baffling the situation; in the instance of his changed behavior after visiting his father, she reports that "much

I study'd," using a significant word, and "I reflected on all Things, from the Beginning to the End, but could find nothing whereof to accuse myself" (28). Ultimately, her only recourse is to consider Bosvil something outside the realm of the knowable: "I endeavour'd to be resign'd, and bring my Thoughts and Inclinations to a true Submission to the Will of Heaven" (29). Galesia pronounces that "Love is like Ghosts or Spirits" (33), an anticipation of the assertion of the importance and impenetrability of the unknown that will occupy so much of *The Lining of the Patch-Work Screen.*

Bosvil's self as it comes to Galesia through the evidence or interpretation of others is as confusing as what she perceives herself. When mutual friends observe Bosvil's partiality for Galesia early in the narrative and ask Bosvil about it, he announces that he is "fixed on my Neighbour *Lowland's* Daughter, and hope shortly to enjoy your good Company, with the rest of my Friends and Relations, at the celebration of our Marriage," an "Answer my Friend little expected to receive," Galesia reports (*LI*, 19). But several weeks later, without explanation, he professes his undying love for Galesia and proposes marriage. In the end, after she sends him a pair of horns as a wedding gift, Bosvil avoids Galesia "almost to the Breach of common Civility, by which I fancy'd I was the Object of his Aversion," a reasonable conclusion but one contradicted by a "Confidant" of Bosvil's who "assur'd me of the contrary, and that Bosvil had told him, That Love had taken such deep root in his Soul, that in Spight of all his Efforts, even Marriage it self, he could not eradicate it, and therefore avoided my Presence, because he cou'd not see me with Indifference" (44–45). This Confidant then provides explanations for Bosvil's behavior throughout the narrative— the story of being engaged to Miss Lowland, the refusal to see Galesia when he recovers from severe illness, and so on—but Galesia doubts these explanations. "How far this was sincere or pretended, I know not, but I rather think he set it up as a Screen to his own Falshood," she concludes (45). Since the novel ends with Galesia's inconclusive thoughts on his behavior ("I know not," "I rather think") and not a reliable third-party explanation, Barker is leaving her readers in a similar state: Bosvil's true thoughts and feelings will never be known and comprehended by either Galesia or the audience, nor are they offered as such.

Barker also raises the possibility that Bosvil does not fully understand Galesia. He admits that he depends upon her appearance: her blush and sighs "made me know, (that, young as you were) you understood that Language" although at the same time he maintains that Galesia's "reserv'd Mein" has "often deter'd" him from declaring his passion (*LI*, 23, 22). Galesia herself takes pains to hide the true state of her feelings, never putting into words her feelings for him as the codes of female modesty require, and hiding her jealousy or rage on several occasions under the veneer of the "Civilities of the Occasion" (38). When their friend Towrissa reconciles them upon Bosvil's engagement to another woman, Galesia says they "planted the Batteries of our Eyes against each others Hearts" but then she gives a "feign'd Laugh, to stifle a real Sigh" and offers "feign'd Smiles in my Face, and jocose Words in my Mouth" (38–39). After he is married, "I was forc'd to act the Part of patient *Grizel*," Galesia explains, and "perform all the farce of a well pleas'd Kinswoman" (45–46). Despite these appearances, Galesia is certain that Bosvil understands the true state of her feelings just as he claims to be certain that she understands his. Although "it was with great Difficulty that I restrain'd my foolish Tongue from telling the Fondness of my Heart," Galesia explains, "the Restraint was with such broken Words, stoln Sighs, suppress'd Tears, that the merest Fresh-man in Love's Academy could not but read and understand that Language, much more he that had pass'd Graduate amongst the Town-Amours" (24). Nevertheless, because there is so much that Galesia admits that she does not understand or know, her own assertions of what Bosvil knew or must have known cannot be relied upon. Whatever claims to certitude each makes for him- or herself or for the other, the third character in this relationship is the unknown; regardless of whether the reader feels certain enough about either character to determine why the relationship ended badly, the narrative emphasizes that these characters remain enigmatical to each other because their ability to collect and process data is insufficient.

Confusion is not the overriding characteristic of human existence, however, at least not according to *Love Intrigues*. As chapter 1 discusses, for someone like Behn the instability of the self precluded its ability to understand anything external or internal, but that is not the case for narratives like Barker's. Just as Galesia achieves a profound understanding of aspects of the natural world, either through her own observation or through careful study of texts such as William Harvey's *On the Motion of the Heart and Blood in Animals* (1628), so Galesia studies and learns her own feelings. Significantly, however, Galesia achieves understanding through retrospection: "too late," she laments at one point, "I found my Folly and Weakness in this my opinionated Wisdom" (*LI*, 17). She also occasionally retells how she gained a deeper understanding of her feelings as events unfold.

Upon learning that Bosvil really is engaged, Galesia opposes her feelings with what she thought would be her feelings: "Behold now, my *Lucasia,* what was to become of all my Resolutions and fancy'd Indifferency," Galesia explains, mocking the "fancy'd Satisfaction I took in his Absence" and discovering "so these my Resolutions were all meer Fantomes, compos'd of Vapours, and carry'd about with Fancy, and next Day reduc'd to nothing" (39). Galesia repeats "fancy'd" to emphasize the discovered delusion, drawing on the vocabulary of the supernatural—"Fantomes" and "Vapours"—to emphasize the initial error and final true understanding.

Overall, Galesia's understanding of her emotional state during the conflict with Bosvil is more profound after studying it from the distance of time than it is during events. This difference suggests that while the self may not always be wholly penetrable to its owner at any given moment, study from a more detached perspective does yield deeper understanding, not something Galesia (or the reader) can achieve with Bosvil. Furthermore, such knowledge does not require a stable self, but can encompass a changeable self, even an irrational one. Galesia readily acknowledges when she is not driven by reason. "Madness and rage seiz'd me, Revenge and Malice was all I thought upon; inspir'd by an evil Genius, I resolv'd [Bosvil's] Death, and pleas'd myself in the Fancy of a barbarous Revenge, and delighted myself to think I saw his Blood pour out of his false Heart," she explains when she learns that Bosvil has proposed another man for her husband after proposing himself to her, calling such notions "wild Thoughts" and emphasizing their irrationality with "inspir'd by an evil Genius" and "Fancy" (31–32). The unknown means others, but it does not mean oneself.

Other texts also investigate this question of knowing oneself but not the selves, or larger world, outside that self. Eliza Haywood's *British Recluse* (1722), for example, shares these concerns and this structure for exploring it. Haywood's amatory narrative demonstrates how female authors could see these are particular pressing issues for women: understanding of self and other, usually in the form of the unknown and therefore dangerous lover, could be considered a fundamental issue for amatory fiction, after all. As Kathleen Lubey argues, Haywood's vivid depiction of her characters' emotional states is meant to teach readers about their inner lives by representing emotional states and by generating those states in readers: "The purpose of her fiction, Haywood asserts, is to improve readers' capacity to reflect on their interiors and experiences.... Readers must 'be sensible' of—that is, both aroused by *and* detached from—their own passionate 'falling' into the immoderate states of excess about which they read."[30] Like *Love Intrigues*, Haywood's novel involves a young woman telling the story of how she was mistaken in her lover, and how her inability to know him and his motives led to disappointment and difficulty. Haywood complicates the issue with a doubled plot: there are two protagonists, Belinda and Cleomira, and each tells her own story of love disappointed in this way. Also like *Love Intrigues*, *The British Recluse* explores the possibility of self-knowledge even when the self is changeable, perhaps unpredictable, and certainly irrational and passionate. Describing her reaction when Lysander first approaches her at a ball, Cleomira explains, "Surprise, and Joy, and Hope, and Fear, and Shame, at once assaulted me and hurried my wild Spirits with such Vehemence that had I answered at all, it must have been something strangely Incoherent."[31] Belinda recognizes that when Courtal failed to declare himself despite signs of his attraction to her, "My Fears now turned to Indignation! I raged to think my Wishes had deceived me! and half despised him for his Insensibility!" (*BR*, 206) Like Galesia, despite not following the social rules governing female behavior, Cleomira and Belinda are fully aware of how they feel, why they feel it, and what those feelings make them think and do.

Similarly, Haywood's novel shows the difference between knowing oneself and knowing what is going on beyond it. Haywood's young women might know themselves but they certainly do not understand the reality of the young men to whom they are attracted, and this ignorance is a crucial factor in their actions and ultimate humiliation. Unable to penetrate the façade which Lysander and Courtal each affect, both young women accept highly conventionalized speeches of romantic love and physical attractions as reality; because they expect to get what they think they have gotten, they are vulnerable to an external unknown. Lysander seduces Cleomira, impregnates her, abandons her, and marries another. She is astonished repeatedly by his oscillating attention and neglect until evidence of his absolute rejection comes in the form of the actual marriage ceremony that he celebrates with another woman (*BR*, 194). Belinda does not lose her virginity to Courtal, close as she comes to doing so, but she does lose her fiancé, Worthly, when he duels with Courtal and then, recovering from a supposedly fatal wound received in the duel, marries Belinda's sister instead. Belinda disgraces herself further when she chases Courtal to London, where she discovers that his name is assumed and that he already has a wife, a mistress, and a past full of seductions and abductions. The final revelation arrives near the end of *The British Recluse*, when Cleomira and Belinda discover that Lysander and Courtal are the same man, whatever his name—and he has used quite a number of aliases by the end of the novel—but preserve their resolution to retire from the world together. This structure emphasizes the difference between knowledge of the interior or internal world—their selves—and knowledge (or ignorance) of the exterior or external world—their lovers, their families—by making the narrators so intensively and accurately aware of themselves, and their romantic tragedy so much the result of problems understanding others.

Barker's interest in the limits that the self imposes on knowledge expands beyond the world of women and romantic relationships in the second and third narratives of the Galesia trilogy, however, indicating that although these questions were indeed pressing for women, Barker also recognized or was more prepared to investigate their larger epistemological ramifications. The relatively simple framing device of *Love Intrigues* becomes far more complex and central to the ideas of *A Patch-Work Screen* and *The Lining of the Patch-Work Screen*, morphing into a more ostentatiously layered structure, not only within individual novels of the trilogy but also among the three books of the set. This structure constructs a communal model of knowing and of the self that uses aggregate perception and interpretation to compensate for the limits of the knowable and the limits of the individual knower. James Fitzmaurice tracks Barker's changing attitudes toward learned men between 1688 and 1723, and differences between her 1713 publication and those of the 1720s confirm that her attitudes toward natural philosophy were evolving, although he attributes these changes to Barker's aging and mellowing.[32] Furthermore, if King is correct that Barker had grown isolated personally, religiously, and politically during the previous decade in Wilsthorpe, then it is equally unsurprising to find Barker responding to this isolation in the 1720s by returning to St. Germain—in effect, returning home from exile—and by writing narratives about the importance of communal effort and the problems of a world incomprehensible to the isolated individual.[33]

The timing for these structural and thematic intensifications is suggestive for literary reasons, as well. By the 1720s, natural philosophy and ideas of the self had come to focus increasingly on the individual as the locus of knowledge production or locus of political ideology, which ran hand-in-hand with philosophical developments. This increasing focus was supported and/or reflected by the deification of Isaac Newton as a singular authority. It was also accompanied by an ascendance of Newtonian, mathematical philosophy interested in a dispassionate, quantitative method for describing the cosmos, and a growing fascination with the Lockean self.[34] It is not surprising that Barker, who was interested in the self's difficulty comprehending incoming data, should respond to such developments with narratives advocating a communally grounded epistemology and warning against the dangers of reductive thinking. As chapter 4 and chapter 5 show, this is not a phenomenon limited to Barker in the 1720s. Barker's Galesia novels offer an example of one response to the issues in natural philosophy's idea of the self and the epistemology built on it that became particularly significant during this period, a response possibly intensified by Barker's other interests and experiences.

Barker's last work, *The Lining to the Patch-Work Screen*, is most strongly focused on the limits of natural philosophy and the limits of the knowable. Barker's concern with those boundaries should not be understood as a refutation of natural philosophy. Debates about the limits of the knowable raged throughout the English and Continental philosophical community beginning well before Barker began to write and continuing well after her death. These debates posed key questions—were humans limited in their knowledge because of their selves, their fallible minds and bodies? Were humans limited because of God, because some things humans simply cannot know?[35]—and produced methodologies to address them. Owing to the nature of the questions about nature and humanity, there was considerable overlap between philosophy and theology; as Amos Funkenstein points out, this overlap precipitated a number of "secular theologians," thinkers like Robert Boyle, Robert Hooke, and Isaac Newton who grappled with theological questions from outside clerical structures.[36] Conflicts over whether humans could ever approximate a prelapsarian accuracy of sense and intellect contributed to religious as well as philosophical conflicts from the sixteenth century through the early eighteenth, including both English Protestants and English Catholics like Barker.[37] Arguments about whether there were unknowable realms and if they were unknowable by divine fiat or human fallibility included discussions of miracles, spirits and ghosts, and the interpretation of Scripture. As early as *Love Intrigues* but especially in *The Lining for the Patch-Work Screen*, Barker demands respect for the unknown and insists on recognizing the limits of human capacity in ways that complicate and extend experimental philosophy's ideas about the individual self's relationship with the community.

Barker's signature strategy, her narrative structure, introduces the theme of the knowable and unknowable as early as the dedication to *The Lining of the Patch-Work Screen* by presenting Galesia's departure for London in winter as inexplicable:

> ...some Business of consequence call'd her to *London*, whether masquerading, or Tossing of Coffee-Grounds, I know not; but probably the latter; it being an Augury very much in vogue, and as true, as any by which *Sidrophel* prognosticated, even when he took the Boy's Kite for a blazing Comet; and as useful too as Scates in *Spain*, or Fans in *Moscovy;* whatever was the Motive, our *Galecia* must needs ramble, like others, to take *London*-Air, when it is most substantially to be distinguished, in the midst of Winter.[38]

This passage is full of the unknown: "I know not," the frame narrator reports, emphasizing again later, "whatever was the Motive." The middle section speaks ironically or skeptically about the "Tossing of Coffee-Grounds," "Augury," and other forms of "prognostication," suggesting that at least certain methods for obtaining knowledge yield error when they yield anything at all. The mystery remains to the end of the novel, when Galesia's friend "the good Lady" invites Galesia to return with her to the country and Galesia immediately packs up without explanation: "This Invitation was an inexpressible joy to our *Galecia;* so she hastned to put up everything; the Gentlewoman lending her helping hand; soon finished and took her away in the Chariot to her Inn that night, in order to prosecute their Journey early the next Morning" and there the book ends (*LPWS*, 290). This mysterious motive for Galesia's visit to and departure from London frames the narrative just as her arrival in the dedication and her departure at the end frame the narrative. The book thus begins and ends with the largest, overarching mystery of Galesia's behavior, something we are never told. Galesia's behavior has reasons—the frame narrator points out that there is a "Motive" for Galesia to go to London—but they are not known to the frame narrator or to the reader, only to Galesia.

The supernatural also raises ideas of the unknown and the unknowable in the third novel.[39] Barker invokes the limits of what is knowable or known with the narrative's preoccupation with death, signaled by a remarkable number of widows, but she also spends considerable narrative energy on tales involving the supernatural. Jennifer Frangos points out that death itself was questionable territory for natural philosophers: what side of the limits of empiricism was it on?[40] *The Lining of the Patch-Work Screen* opens with Galesia, having reflected upon her life and the use of maturity and old age, pondering slightly bitterly those who, "When the Blossom of Youth is shed," do not "bring forth the Fruits of good Works" or "assist the Feeble, with other Works of Mercy corporal and spiritual?" (*LPWS*, 181) Barker not only links "corporal" and "spiritual" but also connects those two dimensions of human existence to the next sentence, when the narrative directly introduces a mysterious stranger:

> Do we instruct the Ignorant, comfort the Afflicted, strengthen the Doubtful, or assist the Feeble, with other Works of Mercy corporal and spiritual?
>
> She was thus ruminating, when a Gentleman enter'd the Room, the Door being a jar. He was tall, and stood upright before her; but not speaking a word, though she look'd earnestly upon him, could not call to mind that she knew him, nor could well determine whether he was a Person or a Spectre. At last she ask'd him, who he was; but he gave her no answer. Pray, said she, tell me; if you are a Mortal, speak; still not Answer. At last, with an amazed Voice, she said, pray, tell me, who, or what you are. (*LPWS*, 181)

At this juncture, he tells her that he is her "old Friend Captain *Manly*" and she "was extreamly confused, to think that she had so weak an *Idea* of so good a Friend, as not to know him" (181). This scene suggests that the boundary between what is known and unknown, between the familiar and unfamiliar, the living and the dead, the corporeal and the spiritual, is liminal. The prolonged suspense before Captain Manly reveals who and what he is emphasizes the sense of the unknown and the relief when illumination is possible. The language of this passage also operates on the boundary between the certain and the uncertain. The frame narrator and Galesia both supply the certainty, the former by explaining what is going on and the latter by recounting her thought process. Neither Galesia nor the narrator can know what the mysterious visitor is until he decides to make it known, however; the knowable does not belong to either the frame narrator or the narrating character, but to a third party, external to both and beyond their control. Furthermore, Captain Manly's unsettling visit, so nearly a visitation, sets off a chain of stories involving the supernatural, particularly ghosts and apparitions, in the first section of this novel. Captain Manly may or may not be a visiting specter, and then he tells a story in which another man, another captain, tells a story about being visited by the ghosts of three sailors on his ship, who drowned (199–200). Like Captain Manly, the three sailors appear so they can tell a story and obtain a desired result: what happened to them and what each would like their captain to complete for them. In this regard, they are consistent with traditional ghost stories up to this period, in which ghosts often return to set something right.[41]

Barker's position on the supernatural was consistent with English natural philosophy's interest in that realm and more complicated than a rejection of Cartesian materialism, Lockean empiricism, and Baconian principles.[42] Stories of ghosts and other supernatural phenomena—such as *A True Report of Mrs. Veal*, for example, rather than the obviously fictional tales of incorporeal visitants that proliferated from the early nineteenth century on—could be "empirical proof of God's continued intervention in the world," as they potentially offered insight or grounds for investigation into the immaterial aspects of God's creation.[43] As Troy Boone points out, authors of apparition narratives could accept and question rationalism at the same time.[44] Spirits were also part of debates about the nature of matter. Philosophers as different as Isaac Newton, Robert Boyle, and Anne Conway all pondered the question of whether spirit occupied middle ground between the material and the immaterial, and spirit remained part of the debate about how matter—including the human body—was animated.[45] Debates over materialism thus kept discussion of "spirit" in a wide variety of senses alive during the late seventeenth century; in particular, "spirit" became an important concept for retaining a role for the divine in natural philosophy.[46] The use of ghosts and spirits in these ways signals Barker's involvement in natural philosophy, specifically in a theologically infused natural philosophy such as those advocated by Robert Boyle or Isaac Newton, rather than an atomistic natural philosophy consistent with Hobbesian or Lucretian thought. The function as well as the content of these stories also helps Barker call attention to the limits of human knowledge as it is bounded by human perception, a key debate in philosophical circles.

As if the ghost story is not enough to remind readers that there is a realm beyond that which can be known, a supernatural realm, Barker makes this story about knowing—the ghosts return only to provide information—and emphasizes the direct knowledge that the original teller had of the story. Captain Manly observes in recounting this story for Galesia, "tho' we hear many Stories of Spirits and Apparitions, and greatly attested for Truth; yet we seldom meet with any body that can relate them of their own knowledge, as did this Captain" (*LPWS*, 200). The investigation of ghosts, spirits, and apparitions by empiricists, including the Deists and the Platonists who investigated stories of the supernatural during the late seventeenth and early eighteenth century were attempts either to explain the seeming suspensions of the laws of nature in philosophical terms, or to use those seeming suspensions as evidence of areas whose laws could be discovered.[47] Barker's self-conscious recognition that ghost stories are not usually credible because they are never at first-hand is not a rejection of the idea that there are ghosts or spirits—Galesia seriously asks if Captain Manly is a "spectre," after all—but an evaluation of the empirical methods used to investigate them, including the conversion of such experiences into philosophical accounts and opportunities for virtual witnessing. Although Captain Manly points out that it is surprising to hear the story from the person who experienced it, suggesting a claim for truthfulness, the novel actually presents the story at the several removes that Captain Manly has just been commenting on: the frame narrator tells the audience about Galesia hearing the story from Captain Manly, who tells the story of his hearing it from the sea captain, a story that includes the actual story that the sea captain tells Captain Manly. Barker detaches this ghost story from the moorings that would have made it unquestionable, eyewitness testimony. It is not an account from a modest witness and as a result, it is suffused with dubiety. The tale itself is about, and surrounded by, epistemological indeterminacy.

The third encounter with the supernatural heightens the attention on the difference between kinds and degrees of knowledge. After seeing Captain Manly to the door, Galesia reads a story indebted to a variety of sources, most clearly Chaucer's "Miller's Tale" in *The Canterbury Tales* and the French fabliau "The Corpse Twice Killed" (*LPWS*, 202–5).[48] In Barker's version of this story, a Knight inadvertently beats to death the novice who attempted to seduce the Knight's wife, then tosses the body over the wall into the neighboring monastery from which the novice has come. The Knight and a second novice from the neighboring monastery proceed to push the body back and forth over the wall, each man thinking—only one rightly—that he has killed the dead man. The second novice, who had long feuded with the dead man and mistakenly believed that he had killed him, experiences a series of coincidences and mishaps in which he cannot escape the body of his dead colleague. The second novice is convinced that the ghost of the dead man is chasing him and makes exclamations such as, "[I]f nothing will appease thy Ghost by my Blood, I am ready to resign my Life to the Stroke of Justice" (*LPWS*, 205). Eventually everyone is released from the literal and metaphorical prison of confusion and deception by enlightenment: the true killer of the novice confesses, the king and monastery pardon everyone, the dead man is buried, and the rest of the players live happily ever after. The comedy here lies in the difference between true and false knowledge, between the conclusions drawn from a series of events interpreted through emotion (the second novice's conviction that the stone he threw killed the man he detested, for example) and those drawn through full and complete knowledge. While the second novice is convinced that the body or ghost of his friend is chasing him, for example, the Knight has no such concerns; he just wants to get rid of the body. As a result, the novice is the one in trouble by the end of the story while the Knight shows every sign of getting away with murder. Only the Knight's refusal to let an innocent man be executed can change the course of events and the admission of his guilt poses no threat to his own well-being, as his rank and relationship with his King secure his pardon. The Knight's understanding of the situation, from the secret of the novice's infatuation with his wife to the secret of his own culpability, highlight the flaws in the frantic second novice's perspective and suggest that the stakes in understanding the external world are quite high.

This series of encounters with the supernatural—Captain Manly's appearance, the ghost story he tells, and the tale of the dead novice that Galesia reads after he leaves—offers a sequential commentary on the limits of knowability, from the explicable, natural, and respectable; through the inexplicable, truly supernatural, and respectable; to the seemingly inexplicable, falsely supernatural, and laughable. The truly unknowable exists, according to Barker, and it deserves respect, but it must be distinguished from the known by careful, persistent inquiry open to all explanations, even a supernatural one. Credulity is not acceptable, but one must still recognize the limits of human capacity. Barker's sequential structure is also a bookended structure. Captain Manly's unnerving appearance and Barker's version of the "Corpse Twice Killed" are stories pointing to the falsely supernatural. The tale of truly supernatural events—the ghostly appearance of the three dead sailors—sits between these stories, to contrast with the inaccuracy and to demonstrate healthy skepticism at work. This trio, both as a sequence and as a pair of stories flanking a third tale inverting their primary conceit, thus demonstrates the limits of human perception and the role of the self in understanding external data.

In "The Cause of the Moors Overrunning Spain," a story within a story in *The Lining of the Patch-Work Screen* that follows the three ghost stories, Barker expands the use of the supernatural. On one level,

Barker uses this story to critique the relationship between politics and empirical philosophy. At the start of this story, which Galesia reads in a book, the King of Spain dies, leaving his "little Son" in the care of the late King's brother (*LPWS*, 205). Predictably, the regent usurps the throne, sending the widowed queen and her son to the Moors for an army with which to invade Spain. In the course of the war between the usurping King's Spanish army and the Moorish King's army, the usurper rapes the daughter of his best general, causing the general to revolt. That rebellion in turn drives the usurping King to seek the Devil's help in the Devil's Tower, a mysterious tower whose origins were unknown (*LPWS*, 207–208). At this point, the story of the "Moors Overrunning Spain" changes from a fairly conventional tragic-heroic romance to something far less classifiable. The usurper and his followers use procedures and logic straight out of experimental philosophy to cope with the supernatural beasts and phenomena they encounter in the Devil's Tower. When the Devil's Tower is opened "with great difficulty," the people who walk in "were immediately suffocated, and fell down dead; which was surprising at first; but on second thoughts, it was easily concluded to be the unwholesome Vapours, so long shut up from Air, which caus'd that sudden Stop of the vital Spirits," an amazingly clinical conclusion given that they are opening something called the Devil's Tower in search of a consultation with the Devil himself (208). Applying a knowledge of the behavior of gasses, the company decides to "let it stand open a few Days" to let the noxious air out, and they procure torches "not only for the Service of their Light, but to help extenuate those poysonous Particles there gather'd by means of the want of Air," a passage that concludes with the reminder that "Thus they entered the Habitation of the Devil" (208).

Their journey into the Devil's Tower is replete with supernatural and unnatural sights and sounds, to which they respond as they did to the sudden expiration of the first entrants. When they encounter a "vast fiery Furnace, wherein were many Monsters marching about," the usurping King and his band try to "conceive the true natural Cause of these Productions, whether a subterraneous Fire heated that Red Liquor, which appear'd like Blood, (which Liquor, perhaps, was only Water, so coloured by passing through Red Earth) no body could conclude; tho' every one made their several Conjectures thereon" (*LPWS*, 209). In the end, the usurping King and his party encounter two riddles, one engraved on a "Gate" in the Devil's Tower leading into the base of the Tower, the other on a "great Image of TIME" occupying the center of this space, each riddle warning against "looking behind"

(209–10). Although the King and his party attempt to decipher the riddles, they eventually settle for the most literal interpretation: not to cast their gaze behind the "great Image of TIME" that stands in the center of the round room at the bottom of the Devil's Tower. After they retreat from the Tower, it sinks into the ground over night and the Moors "over ran" the "Country for many Years after" (210). "Thus," the story warns, "the Devil's Oracles are always double and delusive, and such are all his Temptations" (210).

Barker's image of the usurping King and his group gingerly advancing through the Devil's precincts, attempting to use reason to make scientific sense of monsters, giant mills grinding "Human Creatures" as their grist, a "vast Cauldron full of Blood, which kept continually boiling, but no Fire was to be perceived," and so on (*LPWS*, 208–9) quickly becomes ludicrous. If it is just conceivable that a King and his followers, deliberately opening something called the Devil's Tower in search of the Devil's help, might rationalize the striking dead of people going into the tower, it is considerably less conceivable that when confronted with monsters about to put some of their fellows into the cauldron of blood, this group would pause to argue about whether it only "appear'd like Blood" and whether it is heated by diabolical powers or a "subterraneous Fire" (208–9). These people are ridiculous, and this ridiculousness confirms and is confirmed by the fact that they are led by a usurper and a rapist.

Aside from the dark comedy in this story, Barker offers a cautionary tale about problems of reading and inquiring. This group appears foolish because they cannot reconcile the rational and the supernatural, because having pursued what is beyond the natural, they cannot then accept that different laws might apply. Barker puts two sets of words in conflict: "thoughts," "mind," "conceive," "Conjectures," and "Curiosity" against "perceived," "Appearances," "seem'd," and the phrase "they could not tell" (*LPWS*, 208–9). The problem for the usurper's party is that what they perceive does not reconcile with the powers of their minds—according to the laws of nature, a bubbling cauldron must be heated by something, for example, and human beings are not really the grist for mills. Nobody "could...conceive the true natural Cause of these Productions," Barker suggests, because there is no true natural cause: it is all supernaturally caused. It is the Devil's tower, after all. The usurping King and his party attempt to apply the rules for perceiving and interpreting, that is, philosophical method, for the natural world, for God's world, on the supernatural world, on the Devil's world. They follow the Lockean program, "literalizing the metaphor," in McKeon's words, and "subjecting revelation to material tests of veracity."[49] Certainly in reading the riddle of looking "behind Time" as literally looking behind themselves or the statue, the King's group fails to grasp the more abstract truths of the warning. Ultimately, without recognizing the limits of their method, particularly the ways in which it is not universally applicable, the usurping King and his followers are doomed to failure. Whatever the benefits of the experimental method, they are limited to certain times, places, and subjects. The method is not, Barker suggests, universally applicable. Something else needs to fill the void.

Like Behn, Barker recognized the political implications of the epistemology she interrogated. After all, politics, religion, and philosophy overlapped to a considerable degree during this period. When the usurping King fails to make sense of what he discovers in the Devil's Tower, he leaves Spain open to the Moors (*LPWS*, 210). While the story certainly does not endorse this character, it is also ambivalent about the triumph of the Moors who, although enlisted by the rightful heir, are not Christian and not fighting for the Christian heir but for themselves. Tied overtly to the fall of a Christian monarchy, the usurping King's failure to recognize what is unknowable, to accept that some things are inexplicable, and to live as if the universe contained both, connects the blindness of too much rationalism with a failure of political integrity. Leigh Eicke points out that an inability to interpret lies at the heart of this story, reading it as a political fable about the fall of James II.[50] The failure of interpretation is also overtly connected to empiricism and rationalism, adding another discourse to the commentary. The usurping King and his party misread what they see, whether it is a bubbling cauldron or a riddle inscribed on a statue. If Protestantism's approach to text—in which everything, as McKeon puts it, is a "great sign system"—enabled new epistemologies, then Barker's scene of empiricists reading this way to their destruction is a rejection of this mode of reading.[51] Some things simply cannot be understood. In telling the story of the fall of Christian Spain in terms taken directly from empirical philosophy, Barker associates rationalism, civilization, Anglicanism, and the Revolution of 1688 to create a narrative of usurpation, a failure of religion, morality and ethics. A value for the immaterial—loyalty, faith, ideals—does not threaten civilization, but in fact holds it together.

At that same time that Barker more generally and Haywood in a more limited arena depict the impossibility of reliably knowing others' selves, they also construct novels in which multiple and shifting points of view that present the different selves together compensate for those limitations in order to present a more complete story and a profounder understanding of the issues than a single point of view could provide. Haywood and Barker use the same basic structure, modifying it as necessary: a frame narrator for the novel as a whole, and at least one character narrator who tells the story of herself and its limited ability to know the world reliably. The frame narrator opens and closes these narratives, but she also appears throughout. At such moments, the frame narrator remains outside the stream of the character narrator's account in order to present material that the character narrator cannot, such as what two characters are thinking in response to the same event, or what the character narrator is doing, not just feeling. Haywood's *British Recluse* opens like a Behn story, with a philosophical rumination in the first person: "Of all the *Foibles* Youth and Inexperience is liable to fall into, there is none, I think, of more dangerous Consequence, then too easily giving Credit to what we hear" (*BR*, 155). This narrator testifies to the truth of what will follow—"the following little History (which I can affirm for Truth, having it from the Mouths of those chiefly concerned in it)" (155)—introduces Belinda, "a young Lady of considerable fortune in *Warwickshire*" (155)—and carries the story until the landlady tells Belinda the story of Cleomira's arrival at the boarding house, quickly followed by the dual-narrative sub-structure of Cleomira and then Belinda each recounting her own story. Barker's *Love Intrigues* also begins with a philosophical rumination, but not with the overt declaration of authenticity; details of physical, political, and emotional setting do that work instead (*LI*, 7). Barker complicates this structure in the second two novels. They eschew the ruminative opening to focus on the narrator who knows Galesia—"When we parted from *Galesia* last, it was in St. *Germain's Garden;* and now we meet with her in *England,* travelling in a Stage-Coach from *London* Northward," she says to open *A Patch-Work Screen for the Ladies* (*PWS*, 55), for example—and they present a large number of character narrators, some speaking at the same narrative level as Galesia (she tells them a story, they tell her a story) and others at a deeper level (a character tells a story in which another character tells a story).

This structural distinction between frame narrator and character narrator enables authors like Barker and Haywood to avoid the problem of representing "someone else's reality." With each character responsible for his or her own subjective life, no narratorial authority can be doubted. The narrator simply reports on what she can be expected to know: who did what, that is, the external data, and the moral framework, which she knows because it is her story, after all. Such a maneuver neatly sidesteps the problem of paralepsis, that is, of a first-person narrator knowing more than the laws of physics permit. Or, as Reudiger Heinze puts, it, a narrator "whose quantitative and qualitative knowledge about events, other characters, etc., clearly exceeds what one could expect of a human consciousness and would thus make them prone to being labeled 'omniscient.'"[52] For authors of the early novel, then, one solution for representing knowledge without making unbelievable claims for the powers of the self was to honor the limitations of the self, and to unite as many limited selves as possible in order to present as complete a picture as possible.

In addition to avoiding the problem of presenting an impossible self, this kind of layered narrative also allows authors to explore the questions of knowability and the self that make such a knowledgeable self "impossible." As the titles of the second and third novels in the Galesia series indicate, meaning and particularly the moral core of the novel is a composite, an aggregate of contributions made by the different characters that is compiled by a narrator. Unlike the nearly invisible "editor" of epistolary fiction, this discursive narrator plays an overt organizing role, but the novel itself functions not unlike an epistolary novel, in which meaning is made by the interaction of the letters and the represented subjectivities that are those letters' authors. The patch-work conceit thus applies not just to the novels as a montage of stories which combined make a whole, but also to the nestedness of the novels, that is, the structure of narratives within narratives. Collected, the perspectives make a comprehensive perspective, each offering a different set of interpretations and data that make a whole when assembled.

Barker's use of a layered structure, in part through those perspectival shifts from character to character as well as through a character's shifts from functioning as a character to functioning as a narrator, appears increasingly complexly in the Galesia trilogy. The change in function and therefore in knowledge can be as simple as a character narrating a story to Galesia, who changes position from character narrator to frame narrator for that episode, as in the episode in *Patch-Work Screen* when a visiting Nurse faints when a coach goes by and then tells her story of disappointed love and fortune to a listening Galesia (*PWS*, 119–22). Galesia has been telling her life story to "her ladyship," but when she tells her auditor the story of the visiting Nurse's misfortunes, Galesia shifts position from narrator of her own story to auditor of someone else's with "To which she reply'd" (120). *The Lining of the Patch-Work Screen* is even more complicated, however, as a host of minor characters not only tell their own stories, but tell others' stories and include others telling their own stories within stories, such as Captain Manly's ghost story in *The Lining of the Patch-Work Screen*. The frame narrator, who has introduced Galesia and explained that Galesia's reasons for being in London are unknown, tells the story of Captain Manly telling the story of his life to Galesia, but at a certain point, Captain Manly assumes the voice of the sea captain telling the story of his ghostly sailors. It is not Captain Manly telling about hearing the sea captain tell the story: the first-person speaker stops being Captain Manly and becomes instead the sea captain. In all these novels, other people also become the narrator by telling a story, and the person who had been the narrator moves, temporarily, from being primarily discursive, in Gérard Genette's terms, to being primarily narrative.[53] When this happens, the narrator who had just been presenting his or her own internal, subjective understanding becomes the narrator presenting an external understanding—Captain Manly repeats the sea captain's story but he is not the sea captain, as he reminds us and Galesia when he comments on hearing the story from the person to whom it happened—and each narrator takes a step back from internal, emotional knowledge toward external, factual knowledge. No one entity is responsible for all perspectives but instead, each person is responsible for his or her own perspective. Taken together, they create a complete and authentic representation of internal and external events that can approximate a complete understanding whose biases are compensated for by number, much as the Royal Society compensated for the individual bias of a witness by using a group of witnesses to compensate for the body of the witness.

The relationship of the different narratives within the whole that is *Patch-Work Screen* or *Lining of the Patch-Work Screen* also creates a composite of perspectives that permit the investigation of selfhood and the establishment of reliable knowledge. The titles of Barker's second and third Galesia novels signal that they are built on the idea that a set of seemingly discrete narratives constitute a whole. In *A Patch-Work Screen*, those narratives are held together by a set of techniques including the overarching narrative that is someone telling Galesia's story; the sub-overarching device of Galesia telling stories; the interpolated poetry, which links narratives; and the metaphor of the patches on a screen. In *The Lining to the Patch-Work Screen*, those narratives are held together by a slightly different set of devices, such as a group of recurring triggers—someone telling his or her own story, someone telling the story of someone close by, or a story from a book that is lying in the room—and connective devices or themes. A number of tales are associated with the Moors conquering Spain, for example, but connective devices or themes can also be sequential or can frame interpolated episodes themselves. For example, a traveling Englishwoman tells two stories, the first a version of the tragedy of the Portuguese Nun that she claims to have heard "in the Coach between *Paris* and *Callis*," the second a comedy about a gentlewoman who becomes a gypsy and then a servant until restored to her original class and married to the man of her choosing, a story that the Englishwoman claims to have heard "in the Coach between *Dover* and Home" (*LPWS*, 222–26, 227–38). These stories serve as companion pieces. They follow one another but their origin—the first in a coach on the French side of the Channel, the second in a coach on the English side of the Channel—as well as their theme indicate complementarity rather than sequence. Like the trio of apparition stories catalyzed by the spectral arrival of Captain Manly, these stories are significant both when read in sequence and when read as bookends flanking a connecting episode. In the version of the Portuguese Nun recounted in *Lining to the Patch-Work Screen*, the protagonist rejects her parents—they wish her to leave the convent and join the world—to take up a spurious identity as a nun; abandons that mistaken identity in order to live with her lover by claiming to be his wife although by now she has become a "Bride of Christ" by taking her vows; loses her husband and thus loses her false identity, committing suicide and dooming her children to bastardy and indigence. In the "History of the Lady Gypsie," the protagonist rejects her parents—they wish her to marry a rich, old man—to take up a spurious identity as a gypsy; abandons that mistaken identity to become a servant despite being born a gentlewoman; is on the brink of not being able to marry her beloved when she loses her false identity by being exposed as a gentleman's daughter, taking again her identity as a gentlewoman, marrying her beloved, and becoming the perfect wife.[54] This pair of stories allows Barker to reflect on issues of class, identity, and authenticity, and to do so using similarity as well as difference. The interjected scene, in which the English traveler's auditors are depressed by the first story and nearly go home, serves as a moral gauge: these people have responded correctly, and may be affirmed and rewarded with a similar cautionary tale, only one that ends with success.

Jane Barker's epistemology is not Lockean epistemology, with its emphasis on knowing as an individual act, an individual undertaking that allows the single being to enter into a community of single beings who each has come to know the same things. Nor is it exactly early experimental philosophy's dynamic between individual and group, where the single investigator brings findings to be corroborated by a group of witnesses, after which those findings enter the register of knowledge, a dynamic replaced over the course of the eighteenth century with the Lockean and Newtonian emphasis on the individual. Since then, as Elizabeth Potter puts it, "Modern philosophy has been deeply committed to epistemological individualism, the assumption that the individual is the source of and principle agent in the production of knowledge."[55] Reliability and authority in knowing are thus inextricably tied to singleness and detachment, whether it is detachment from that which is observed or from others who might influence the conclusions drawn from the observations.

Barker's narratives propose a different form of knowing, a different form of gathering knowledge. The layered narratives present the limitations of any individual knower: Galesia can never know Bosvil; even when she collects information from others, her consciousness is the primary one through which all that information and perception is filtered and it can only produce an incomplete understanding. The framing structure also presents the power of the combined knowers, of the community of knowers whose information and perceptions taken in aggregate create understanding, or as much understanding as humans are likely to achieve. Anticipating the objections to the positivist and rationalist epistemologies that arose from Lockean empiricism and its marriage to experimental philosophy, Barker's narratives contend, in Lynn Hankinson Nelson's words, that "communities are the primary epistemological agents." As if describing Barker's community-oriented knowledge systems, Nelson explains that it is the "collaborators, the consensus achievers, and, in more general terms, the agents who generate knowledge are communities and subcommunities, not individuals," contending that the "knowing we do as individuals is derivative, that your knowing or mine depends on *our* knowing, for some 'we.'"[56] Authority and completeness can be achieved as far as the laws of the natural and supernatural worlds allow, but only provided we use a community of observers for seeing.

Rivka Swenson points out that *A Patch-Work Screen for the Ladies* opens by requiring its many different kinds of readers to assemble the different patches, and to add their own, in order to understand not just the novel, but England. Examining the letter that opens *A Patch-Work Screen*, Swenson explains,

> The letter also implies that the patch-works' purposeful obscurity is part of their strength. The letter claims that patch-works reproduce the process by which many readers, 'Whigs *and* Tories, High Church *and* Low Church, Jacobites *and* Williamites,' come together into one

'Fabrick of the UNIVERSE.' The readers who make up the fabric are
required to do their part in making sense of how they, like the narrative that describes them, fit together in history, zooming in on particulars and then back out to consider the big picture. Instead of having the
dots connected for them as they might be in a history-at-large, readers
get a jumble of particulars from which they must create meaning.[57]

Swenson is making a point about Barker's commentary on genre,
but her formulation helps address the possibility of an overarching
consciousness implicit in Barker's structure. The framed narratives
are ultimately contained within one frame, after all, and by using
her own pen name for her protagonist Barker draws attention to the
idea that an author's consciousness has shaped each work. In this
regard, Barker's multifaceted texts approach omniscience—the fly's
eye is, in the end, one eye. This structural unity provided by the first/
last frame is in tension not only with the structure it contains, however, but also with Barker's maneuvers at the end of each narrative
to leave them open-ended. In *Love Intrigues*, for example, Galesia
wonders if she did the right thing and concludes that one must have
faith in God and Providence (46–47); *The Lining of the Patch-Work
Screen* concludes with the phrase "in order to prosecute their Journey
early the next Morning," suggesting further adventures yet unknown
(290). The overarching point of view is singular but significantly, her
knowledge is a patchwork. Each work ends denying the possibility of
complete knowledge, reiterating the impossibility of achieving it with
the same narrative device that Barker uses to propose combining different knowledges to ameliorate that inevitable failure. One person
may be doing the telling but many people have made it possible for
her to know what she says. Swenson's point that the narratives overtly
require investment from the audience offers another argument against
not just closure, but the idea of an overarching, reliable, fully knowledgeable self.

At stake in the emergence of the novel is the definition of the self
and the modes by which that definition is represented and expressed
in text. Barker's opposition to the notions of the self and the conventions by which those notions were being expressed is an opposition
to the idea of a unified, stable, all-seeing self that can understand not
just nature, but other selves as well. In presenting narratives where
full understanding can only be achieved through the contributions
of many and different selves (male, female, educated, uneducated,
fallen, virtuous, Catholic, and so on) and the recognition of the inevitable limits of understanding, Barker and authors like her such as

Eliza Haywood are simultaneously affirming and challenging the values underpinning the philosophical revolution in England and the cultural change it influenced, even as they pioneered the genre that would become the vehicle both for radical and reactionary ideology for the centuries to come.

Chapter Four

The Detached Observer

If there is anyone who epitomizes the ability to remain detached from an experiment and its results, one could make a strong case for Isaac Newton, who once pushed a needle into the socket behind his own eyeball in order to further his inquiries into the nature of optics. Reporting the procedure in his notebook, he writes,

> I took a bodkin (g h) & put it betwixt my eye and ye bone as near to the Backside of my eye as I could: & pressing my eye with ye end of it (soe as to make the curvature a, b c d e f in my eye) there appeared severall white dark & coloured circles r, s, t, etc. Which circles were plainest when I continued to rub my eye with ye point of ye bodkin, but if I held my eye & ye bodkin still, though I continued to presse my eye with it yet ye circles would grow faint & often disappear until I renewed em by moving my eye or ye bodkin.[1]

Consider what it takes to devise such an experiment, then to sharpen the needle, insert it into one's head, press it repeatedly against the back of one's eyeball, and take lucid, thorough notes on the experience. Contrast this detachment with Newton's infamous pugnacity or compare his fierce professional and personal isolation at Cambridge with his behavior starting in 1689, when he became socially and politically active, eventually moving to London to become Warden of the Mint in 1696.[2] By 1703, Newton had maneuvered himself into a government sinecure in a bustling metropolis, secured his half-niece Catherine Barton to run his household and manage his social affairs, and assumed the presidency of the Royal Society, a position he relinquished only at his death in 1727. Nor did he work alone as a philosopher during this time, as his conflict with John Flamsteed attests. Yet despite these thirty years situated within political, economic, familial, and professional networks, Newton was and continues to be understood in terms of isolation. "Solitude," James Gleick pronounces, "was the essential part of his genius."[3]

Newton epitomizes a confluence of ideas about the self during the late seventeenth and early eighteenth centuries. While philosophers increasingly propounded the notion of the autonomous self detachable from its context, they also grappled with the need to establish a community of these autonomous selves and the need to justify their activities through a connection to the larger society. This conflict over the self—was it truly separate from everything else? Could it function effectively and morally if unconnected to nature and other people?—had implications for people outside the philosophical community and engaged the interest of many of them. In previous chapters, I examined the works of authors who were concerned with the nature of the self and its own, individual capacities. In this and the next chapter, I examine the works of authors exploring the relationship between an autonomous self and its larger context, particularly its moral context, in the creation of knowledge.

Eliza Haywood was among the writers and thinkers who recognized the conflict inherent in the notions of self emerging during this period. In *The Tea-Table* (1725), she explores the nature of the self, the role of detachment, and the possibilities for an epistemology that balanced detachment and engagement to promote accurate and moral knowledge and knowing. The relationship between isolation and detachment on one hand and socialization and morality on the other were especially compelling questions for women novelists like Haywood. Barbara M. Benedict argues that the rising value of curiosity in the late seventeenth and early eighteenth centuries created a position from which women novelists could investigate issues that concerned them, such as the nature and role of love or morality.[4] In that respect, detachment offered advantages explored and exploited by Haywood in *The Tea-Table* and in other works. But detachment, both as a quality and as a social position, also posed problems and Haywood explored those as well. *The Tea-Table* uses narrative structure to propose a balance between detachment and incorporation as an epistemological desideratum, suggesting that both qualities and both positions are necessary not just for knowledge but for morally valid and valuable knowledge. Strikingly, the self that Haywood writes in this narrative is disembodied, even as the narrative itself investigates the role of embodiment in balancing engagement and disengagement. Haywood's investigation of the nature of the self thus clearly connects to issues of point of view, suggesting originary techniques and concerns for the development of literary omniscience.

Two issues are central to *The Tea-Table*: the means by which knowledge is acquired and the use-value of that knowledge. Like John Locke's *Essay Concerning Human Understanding*, Haywood's *Tea-Table* suggests that the body contributes to knowledge through sensory input. Joseph Drury points out that also like Locke as well as other philosophers such as Robert Boyle, Haywood afforded the body a role in morality, considering the body's experiences and sensory input significant contributors to the development of a moral sense.[5] Haywood's narrative posits a much more intrusive role for the body in the constitution of the self than Locke's *Essay*, however: for Haywood, the body that one possesses, where that body is located, and how that body is treated by oneself and others are all factors in the creation and dissemination of knowledge, in the ability to know and act morally, and therefore in the quality of self. Haywood's amatory fiction was keenly aware of the dynamic between bodies and minds, whether one's own or someone else's; this issue was a central feature of the genre overall. Her amatory fiction was also aware of how language, particularly narrative, constructed the self and was in turn manipulated by it. The failure of narrative to analyze the situation honestly or to adequately prepare its readers, especially women, concerned Haywood throughout her career.[6] Although *The Tea-Table* is not amatory fiction, it contains an amatory novella and it explores the same questions about the mind and the body that concern many of Haywood's amatory narratives; furthermore, it is a narrative about reading narrative. In *The Tea-Table*, Haywood uses a detached narrator not only to acknowledge the problematic contribution of the body and the value of detachment and separation in acquiring and sharing knowledge but also to advocate for the necessity of social interactions in leavening knowledge with morality.

Haywood's interest in natural philosophy is perhaps most evident later in her career, in the 1740s and 1750s, such as her reference to the longitude competition in *The Invisible Spy* (1755) and essays in *The Female Spectator* (1740–44).[7] Nevertheless, textual and biographical evidence suggest that Haywood was interested in natural philosophy considerably earlier. As a member of Aaron Hill's circle for several years during the 1720s, Haywood was close to Hill, whose enthusiasm for English natural philosophy's idea of applied knowledge for social betterment led him to a number of ill-fated schemes in the years just before establishing his "Hillarian" circle.[8] Aside from guilt by association, however, Haywood's early narratives also suggest an interest in natural philosophy.[9] Drury tracks an interest in Hobbesian materialism through several of Haywood's early novels, including *Love in Excess* (1719) and *The Tea-Table*; Tiffany Potter sees the influence of Hobbes in Haywood's libertines.[10] Haywood translated into English Louis Adrien Duperron de Castera's novel, *La Pierre Philosophale des Dames, ou les Caprices de l'Amour et du*

Destin, Nouvelle Historique (1723), a work that associates alchemy and irrationality.[11]

Images of and references to natural philosophy appear in early novels as well. *Love in Excess* features a scene with Melliora reading Fontenelle.[12] Natural philosophy appears again in the sequel to *The Tea-Table*, *The Tea-Table Part 2* (1726), when Haywood invokes "Philosophy" in her poem, "A Pastoral Dialogue." The speakers, Clorinda and Alexis, alluding to "Athenian Fame" and "Enliven'd Atoms," complain that "When 'tis a Part so tender feels the Wound, / No Comfort in Philosophy is found."[13] Haywood plays with gender around the allusions to "Philosophy" in this poem, as well: Clorinda requests "manly Fortitude," Alexis characterizes women as the "half-Soul'd Sex," and both speakers oppose Clorinda, women, and poetry on one side with Alexis, men, and "Learning" on the other (40). This opposition is consistent with Haywood's profound interest in reason, particularly with the ways in which reason was withheld, either by theorizing or by education, from women. Juliette Merritt contends that "Haywood wishes to advance women's social and personal interests through their capacity to exercise reason," and she groups Haywood with Mary Astell, Catherine Macaulay, and Mary Wollstonecraft.[14] Haywood's well-explored concern with constructions of femininity, with the limits on female power in her society, with problems of agency facing women in her world, is in many ways a concern with the self being described and defined and delineated during her lifetime, a self initially theorized by and still linked to natural philosophy.[15]

The Tea-Table partakes of a significant conflict within natural philosophy over detachment and the self.[16] Admittedly, it is tempting to read *The Tea-Table* in light of Haywood's personal life rather than as an investigation of developing notions of the self and the epistemology built on those notions. The narrative was published in the spring of 1725, several months after Haywood was ejected from Hill's "Hillarian circle," which gathered regularly and often around a tea table.[17] The timing suggests that such a situation and the questions of morality and knowledge that it raises had personal significance for Haywood, and some scholars have viewed *The Tea-Table* as Haywood's bid to get back into Hill's good graces.[18] Haywood's depiction of an idealized tea table, this argument goes, is a quasi-nostalgic recreation of the sociable, erudite discussions of topics both social and literary supposedly undertaken by those gathering at Hill's. As I discuss, however, Haywood's tea table is an investigation of methods for generating knowledge, especially moral knowledge.

Although it can be read as a critique of Hill's tea table in particular, Haywood's analysis ultimately transcends the personal to address the tea table as a model of knowledge generation and to reject literal and metaphorical tea tables centered on male authority. Haywood instead proposes an alternative system, a blending of independent perspective and social discussion performed by men and women that permits disagreement as a route to true and morally true knowledge. In this respect, while it is possible that Haywood's narrative may have begun in catharsis or apologia, it also goes well beyond it. In *The Tea-Table*, Haywood tests the limits of social thinking and individual thinking. She posits a self whose optimal effectiveness in generating and transmitting knowledge is one that balances social separation and social integration.

The *Tea-Table* immediately signals its connections with natural philosophy by opening with a sequence of three passages that invoke issues or conversations generated by the revolution in natural philosophy. The first text is the epigraph, a passage from *The Indian Queen* by Sir Robert Howard and John Dryden that refers to nature being inherently flawed: "*The Heavens have Clouds, and Spots are in the Moon, / A faultless Virtue's to be found in none.*"[19] The allusions to "corrupt" or flawed nature from *The Indian Queen* echo anxieties about nature produced in the wake of Galileo's observations of the moon and other planets expressed by other writers, from John Donne to John Dryden.[20] Haywood's second signal is an "Advertisement" which claims that the representation of society presented in these pages is both flawed and general: "*I have no View to any particular Persons, or Families in the Characters contained in the following Sheets,—The Design of them being only to expose those little Foibles which disgrace Humanity*" (73). While the observations of human nature contained in the narrative are derived from certain characters and are therefore specific, the behavior is general and the conclusions are general truths. Such a claim was standard for philosophers of the period and has a long literary history in exemplarity.[21] The bird you smother in the air pump might be a particular bird in a particular air pump, but it exemplifies a general truth about air, birds (or even living creatures), and air pumps.

The third passage is a rumination on the nature of the tea table as an institution where knowledge is produced. In it, Haywood links *The Tea-Table* to the community of natural philosophers and their methods. The tea table is both an association and a space devoted to the uncovering and sharing of truths possibly hidden behind appearances. Unlike anywhere else, around the tea-table the "Curious" can see things otherwise invisible to those not gathered in that space (73). "Where have the Curious an Opportunity of informing themselves of the Intrigues of the Town, like that they enjoy over a TEA-TABLE, on a Lady's *Visiting-Day*?" the narrator inquires, going on to ask, "Can the Love-sick *Maid*, the Wanton *Wife*, or Amorous *Widow*, be guilty of the least false Step which falls not under the Observance of these *Criticks* in Fame?" (73). The language of this extended passage is striking both for being visually oriented and for claiming powers of sight that produce truth by penetrating appearance. Those gathered around the tea table can see the "Uxorious *Husband*" having an affair with his wife's maid; they can see the "new wedded *Bride*, trembling and blushing at the approach of Night" and know whether "her Modesty be real, or affected?" (73–74). These sights cannot be literally possible for people gathered around the tea table but Haywood collapses the seeing and the knowing into one moment. We learn, further, that this "Cabal" investigates: they "examine," perform a "most strict Scrutiny," and "distinguish" what is "Natural" beauty from that aided by "false hair, fine Cosmetick, or any other Assistant" (73–74).

The combination of visual activity and investigative activity draws an implicit connection to the experimental philosophers of the Royal Society, who also occupied privileged, exclusive space; used that space to examine, scrutinize, and distinguish; and did so to obtain the truth that lay under whatever presented itself to ordinary sight. On its own, the description might not appear to connect to the Royal Society, but situated between two passages invoking and evoking the principles and methods of natural philosophy, this exposition of the nature of the ordinary tea table is charged. Haywood's use of "expatiated" also draws connections to natural philosophy. During this period, "expatiate" as a verb meant "To walk about at large, to roam without restraint; to move about freely in space, wander at will," and although introduced in the mid-sixteenth century, it entered into much more common usage during the seventeenth century with the help of natural philosophers.[22] Haywood's language suggests some criticism of the tea-table investigators, however. They are not only a "Cabal," a loaded word in the highly factional politics of early eighteenth-century England and especially in Walpole's administration, but also "inquisitive Fleerers" and a "rival *Belle*" (73).[23] Now their amazing ability to see what is not before them—the Uxorious Husband with the maid, the blushing bride on her wedding night—seems less prescience and more conjecture. Gathered together, those around the tea table are mean-spirited gossips rather than revealers of reality hidden behind appearance. Without a moral framework, their investigations become "investigations," their insights become "insights."

This description of a tea table signals Haywood's preoccupation with the roles of the community and of morality in the generation of knowledge. Haywood immediately contrasts the fleerers with a community that does have a moral framework and therefore social, critical, and epistemological validity: Amiana's tea table. The hostess's behavior, made prominent by appearing first in the discussion of her virtues, promotes "Wit and good Humour" (74). Amiana has "so much Sweetness of Disposition" that she "inspires[s] the same in others," but also she controls the conversation (74). She does not permit mean talk about absent people and at the same time, shapes their representation:

> None can be Master of a *Virtue* which she does not *Magnifie*; a *Fault*, which she does not carefully *Conceal*, or if too obvious, shadow it over by enhancing the Value of some Perfection.—She excuses the *Whims* of the *Virtuoso* on account of his Philosophy;—the Vanity of the *Poet* for the sake of his good Verses;—the Severity of the *Prude* in respect to that Virtue she assumes;—and the Affectation of the *Coquette*, for the Diversion she affords the Company. (74)

Certain words in this passage reveal the dual nature of the description: "Magnifie," "shadow," and the whimsical "Virtuoso" made foolish "on account of his Philosophy." This vocabulary shows how the tea table is both a tea table and a metaphor for a community that gathers, processes, and produces knowledge. Furthermore, as the visual language—"magnifie," "Conceal," "shadow"—puts the emphasis on what is visible, the passage also acknowledges the manipulation of what is seen, the control of perspective. Seeing, after all, was a central tool and value of the "new science" but as Haywood points out in this passage, "seeing" or "witnessing" is a multi-faceted undertaking.[24] There is no illusion of seeing what is invisible. The Virtuoso, the Prude, the Poet, and the Coquette remain a virtuoso, prude, poet, and coquette, although their social, moral, and epistemological value change: the Poet may be vain, but he is a good poet; the Prude may be severe, but she promotes virtue. Amiana thus seeks to create a community that follows certain moral principles as it generates and transmits knowledge.

Haywood makes consensus crucial to the success of this agenda. Because on one day of the week at least, "none who appear well are refused Admittance," the quality of the discussion is always in peril of degenerating, which is the case in the discussions flanking the main matter of the narrative (75). The coxcomb, hypochondriac, and gossip, presumably admitted because they "appear well," seem more at home at the cabal's tea table than Amiana's. Most days of the week, however, "She endeavours, 'tis certain, to select her Assembly out of such of her Acquaintance, as will not put her to the Blush to defend" (74). While there is a certain suspiciousness to this engineering even selection cannot produce uniformity. Instead, the discussants negotiate and sometimes fail to reach a conclusion, regardless whether they are debating morality or linguistics. Alexander Pettit suggests that "Haywood decenters moral authority and thereby suggests the flexibility of moral categories that she elsewhere presents as rigid," attributing this attitude to Haywood's rejection of a conventional stance in her personal life.[25] There is no moral center but there is morality because everyone subscribes to certain beliefs and practices. The success of the tea table depends not only in part on who is chosen but also in part on the agreement of those individuals to adhere to a certain moral code.

Haywood's critique of the "caballing" tea tables and her rejection of authoritarian tea tables at the start of her narrative are consistent with her indictment in other texts of the socially sanctioned function that a certain kind of tea table served. As Eve Tavor Bannet has pointed out, Haywood's attack on tea tables in *The Female Spectator* is a direct refutation of Addison and Steele's celebration of the tea table as a site of socialization and education.[26] Haywood was profoundly critical of Addison and Steele's celebration of the tea table as a proper site for female activity, as well as for female acquisition of knowledge from men, particularly from Mr. Spectator.[27] For Mr. Spectator, the tea table is a domestic site where a quiet female can imbibe some information along with her tea, but cannot also acquire the skills that would produce that information or for that matter, question and test it independently. Bannet observes that Addison and Steele intended their periodical to educate by providing the news of the day and therefore the subject matter for discussion.[28] For Addison and Steele, the material covered by *The Spectator* was what women needed to know to be fit companions for their husbands. Reading and discussing the paper over the tea table in the morning and afternoon was not just education and not just socialization: it was also socializing education, educating women to be members of society and using socializing to educate them.[29]

In *The Tea-Table* as in *The Female Spectator*, Haywood rejects the idea that the news of the day was the substance on which true education could be built. Haywood also rejects the device of an all-seeing male authority whose vision and prescience create moral and factual knowledge, and the positioning of women as the recipients rather than the generators of knowledge. Although in other texts Haywood might "appropriate, for women's use, the privileges of observation" as Merritt claims, in *The Tea-Table* Haywood suggests that men and women can be possessed either of information or of what passes as it.[30] The information shared by the cabal or by the coxcomb at Amiana's is news but it is hardly edifying. In *The Tea-Table*, the information and insight circulated in the first image of a tea table and in the narrator's first set of discussions at Amiana's tea table are nothing more than gossip, speculation disguised as prescience. As a form of generating or transmitting knowledge, full sociability is a failure, not only for women but also for men and for men and women together. Groupthink in any form simply cannot work.

The Tea-Table is 20 years closer to *The Spectator* than *The Female Spectator*, so it is not surprising to find an early rejection of Addison and Steele's agenda for tea tables here. Haywood's *Tea-Table* also takes on another tea table, one closer to home literally and chronologically: Aaron Hill's tea table, particularly as it was represented by Hill himself in his periodical, *The Plain Dealer*. In No. 11 (27 April 1724), Hill describes the Plain Dealer's visit to a tea table, where he is not integrated into the group at all. Instead, the Plain Dealer is treated like a celebrity by everyone, seated in a special chair, and eagerly solicited for his opinion by the young women in the group:

> When the Tea-Equipage was remov'd, and every One remained silent in Expectation of what I should say, the Charming *Leonilla*, my Friend's Eldest Daughter, after having first fix'd her Eyes on a Young Lady that sat opposite to her, and then on me, desir'd in Behalf of Part of the Company, I would give my Opinion in the Case of *That* Restriction *Young Women lie under, who, from a* Custom, *which, in her Opinion, had no* just Grounds *for its Support, are* forbid *by* Decency, *let their* Passion *be the most* vertuous *and* tender *that was ever felt, to make the* least Advance, *by which the Part belov'd may* discover *the* Sentiments *they have in his Favour.*

In response to this request, the Plain Dealer pronounces upon appropriate female conduct, replying, "*A Woman of Sense can never* feel *that* Tenderness *for a* Fool;—*and a Man of* Sense *would always reckon such a* Declaration, *a* Desire *of* enjoying *that* Happiness, *whose* Basis *is* Love, *and whose* Support *is* Virtue." Having "delivered my Opinion," the Plain Dealer elaborates for the benefit of the other young lady, explaining that "*There was a* false, *as well as a* true

Decency; and that One *consisted in* Grimace *and* Shew, *the* Other *in an* open Deportment, *the* Result *of an unaffected, disinterested* Love of Virtue.... True Virtue *wants no Foil, and is never talking of itself.* False Virtue, *sensible of its* Weakness, *is forc'd to have Recourse to* Hypocrisy, *and is ever boasting a* Strength *it is a* Stranger *to.*" His pronouncement about female virtue triggers a poetic duel between two young men over appropriate behavior for women and men; they in turn provoke a heated discussion until the Plain Dealer's host silences his guests by reminding them that the Plain Dealer will make their conversation public and praises the Plain Dealer for policing people's opinions and expression.[31]

Although Hill provides the content of the discussion, the focus of the narrative never shifts, nor is it supposed to, from the Plain Dealer's view of the disagreement and different opinions. At all times he preserves his thoroughly opinionated, central position and the views of the other speakers—the two young men who read their poems to the company—function to elucidate the Plain Dealer's pronouncement. While Christine Gerrard sees this scene as Hill's ideal community, which indicts "outmoded rituals of female coquetry and male pursuit," it remains a primarily male-oriented and hierarchical space and discussion.[32] The tea table is organized and led by a man (the host), its activities are centered on a man (the Plain Dealer) and reinforced by men (the sparring poets), and its discussion is aimed at regulating female behavior. In personal terms, it is perhaps not surprising in 1725 to find Haywood, however grieved to have lost Hill and his circle, critical of the dynamics of such a community. Given the competition between Haywood and Martha Fowke and the fact that it became highly public through their publications, it is possible that Haywood saw Hill, however beloved, as a highly imperfect regulator of female activity. The connection to later critiques of tea tables speaks to a broader interest, however. Haywood's rejection of conventional tea tables for policing behavior and her recognition of the problems of either a central male figure or a contentious, ill-spirited group as the generator of knowledge have larger implications beyond the biographical.

In Haywood's narrative, the tea table is a sociable place with all the influences and components of society, such as gender, class, and the limitations of good breeding and propriety. Amiana's tea table also is organized and guided by a woman, thereby rejecting male-centered authority in a social gathering, in society, or in a knowledge-system. The information that Amiana's guests exchange is not factual or speculative but philosophical—what is the nature of love? What is the nature of ambition? And so on. They all, women included, practice philosophical skills. While Alexander Pettit complains that Haywood decenters moral authority and that "the *tete-a-tete* is not as harmonious as it first appears. The friends disagree about certain topics, and they leave many of their debates unresolved," in fact this is what makes Amiana's tea table so effective.[33] From Milton's *Areopagitica* to Henry Oldenburg's circulation of information from philosophers around Europe to the disagreements among Fellows about hypothesis and conclusions, disagreement was considered crucial to the advancement of knowledge. As Steven Shapin and Simon Shaffer point out, part of Robert Boyle's achievement in his debate with Thomas Hobbes was establishing that civil, intelligent disagreement was fundamental to the functioning and effectiveness of the new experimental method.[34] Haywood would explore the issue of respectful disagreement between men and women in *The Female Spectator*, as well.[35] So although the conversation is not "harmonious" because characters disagree, it is "harmonious" in the sense that each character's perspective combines with all the others works to create a beautiful whole: inquiry, challenge, and investigation all move deliberately and respectfully toward truth. In collapsing the tea table and ideal philosophical space, Haywood both argues that natural philosophy at its best is like a tea table, and that the best tea tables are like the ideal of natural philosophy—perhaps even the Royal Society.

Nevertheless, Haywood does not suggest that group endeavor is the route to knowledge. Instead, she uses a narrator to balance the contributions of the group in generating morally valid knowledge with the contribution of the independent individual. Haywood calls attention to the narrator through his encounters with the irritating people at Amiana's, his transgression against the group, and his reintegration into it. Haywood establishes him as a person who is both included in but detached from the group in order to make use of that position throughout the rest of the narrative. Unlike the other people gathered around the tea table, he is not invited by Amiana and was not drawn there on his own; in other words, he is neither in her carefully selected, controlled group nor is his attendance motivated by desire or ambition. He has been brought by Dorinthus, who was originally invited by Amiana, and although the narrator claims to become one of the company—he was "sometime ago introduc'd," suggesting that he has since become a more frequent guest at Amiana's tea table, and he counts himself among the five who conduct the conversation (74, 77)—he is not actually a participant. The unnamed narrator, in fact, observes and describes and often concludes, but he does not contribute.

The narrator's detachment during the main portion of the narrative is epitomized by his silence. He is not silent after the fact, of course: he recounts everything that was said with remarkable precision. He also speaks briefly in the conversation before the dialogue that he recounts. During the conversation which he later reports to his own audience, however, he is quite silent. He further recedes when he provides Philetus' narrative within his own narrative (83–85), or when Brilliante reads a "Manuscript Novel" about Beraldus and Celemena (86–103). When he does provide material original to himself, it is quite limited. There is a little interpretation, such as when he calls Brilliante "charming" and "beautiful," or claims that Amiana read a poem "with the most agreeable Voice and Manner in the World" (79, 80, 81). There is also a little narrative information, such as when he says that Dorinthus began speaking after "perceiving [Philetus] had done" or reports that Amiana "had no sooner spoke these Words, than she took the Paper out of her Pocket" (78, 81). In providing such a strict account of what he saw and what he heard, the narrator becomes a witness. In this regard, the narrator is quite different from the other people around Amiana's tea table. He is also quite different from the members of the "Cabal" described at the start of *The Tea-Table*, who carry their perceptions and interpretations even into the realms of the unseen and unheard. And he is quite different from Mr. Spectator, the supposedly silent observer of humanity to whom Haywood's narrator might be compared. Haywood's narrator does not claim to know more than he can see, nor does he claim to see more than he can know, that is, more than is before his eyes. Juliette Merritt argues that in her amatory fiction, Haywood rejected the kind of spectating epitomized by Addison and Steele's Spectator—or by Hill's Plain Dealer—and in *The Tea-Table*, Haywood creates a structure as well as a narrator that rejects it.[36]

With a silent narrator who functions to erase his own presence, the bulk of *The Tea-Table* sometimes reads more like a philosophical dialogue than a narrative. Large sections of it replicate the speeches of other characters. The passions are introduced and defined by Philetus, who is evidently channeling Thomas Hobbes and his *Leviathan*. "*Love*, as it differs from all the other Passions in its Consequences, does so too in the Manner by which it first gains Entrance in the Soul and after wholly engrosses all the Faculties of it," Philetus states, and defines "Ambition" as "that lawless Thirst of Power which inspires in some Mens Breasts such unwarrantable Designs" (78).[37] The narrative then replicates the dialogue verbatim for some time (78–81). Similarly, Brilliante's reading aloud from a "Manuscript Novel" is inserted whole and unedited by the narrator into his own account of the party. When the characters do speak with each other, they respond directly to what is said. After Philetus laments the destructive power of Love, Dorinthus, "perceiving he had done," answers him by acknowledging Philetus' experience—"Though nothing...can be more plain than that you have experienced, to a very great degree, the Power you so well describe"—and then refuting Philetus' claims: "yet I think there may be something added to that you have alledged" (78–79). When Dorinthus finishes, Brilliante says, "Well have you argued, Gentlemen!" but adds, "'Tis not so with us; a thousand different, dread Ideas strike us with Horror at but a Thought of giving up our Honor" (79). Women also answer women: "Your Quotation is extremely just, my Dear!" Amiana applauds when Brilliante finishes her passage from *Jane Shore* and after thanking Brilliante at the conclusion of Beraldus and Celemena, adds that "notwithstanding the Pains you have taken to oblige us, and that there are some lively Strokes of Passion in the Story, I cannot help saying that I think the Character of CELEMENA faulty" (80, 103). Their conversation advances ideas, introducing, testing, modifying and developing them individually and thereby developing the conversation as a fluid whole. As Brilliante observes, "we have insensibly fallen from the particular Passions, to the Temper in general" (86). One might almost say that this is no Boylean experimental narrative, but a record of the conversation following a demonstration or a paper presented to the Royal Society. For a good portion of the text, it is the characters who carry ideas and events along, not the narrator. Their conflicts bring out different ideas and the narrator says almost nothing beyond repeating what happened.

The narrator's silent membership in the group during most of the narrative contrasts with his behavior at the beginning of *The Tea-Table*. In the first stage of events at Amiana's, the narrator converses with a "Titled *Coxcomb*," the "Counterfeit indispos'd," and then "the Lady who found Fault with her" (75–76). The narrator is offended first by the Coxcomb's boorishness and then by the false friend's indiscretion and meanness. He withdraws physically to a window seat and mentally into his own thoughts. "I made but little Answer, being unwilling either to affront, or to say any thing which might look like an Encouragement of a Temper so pernicious to Society," he explains, "and now being desirous of hearing any more of the same Nature, I rose, and, retiring to a distant Window, fell into a deep musing"

(76). In this behavior the narrator anticipates Haywood's celebration of solitary reflection in Book IV of *The Female Spectator*, but here Haywood also does not permit isolation and withdrawal to serve as a solution to the problems of sociability. Instead, this behavior brings the narrator precariously close to transgressing social norms. He is forced to return to the company by Amiana and to apologize by his friend, Dorinthus lest he be cast out from the society he wishes to be part of:

> I became so lost in Thought, that for a good while I was altogether ignorant of what was said or done in the Room; and might perhaps have continued in that *Resvery* [sic] much longer than I did, if the agreeable AMIANA had not called to me, desiring me to join Company, and at the same time, the Person who did me the Favour of introducing me, pluck'd me by the Sleeve, reminding me that the little Impertinencies of those who visited her, ought not to make me forget what was due to a Lady of her Quality and Merits. (76)

Prefacing the quintet's philosophical conversation with such a scene shows how Amiana's best efforts to secure the best company are thwarted by the basics of proper sociability, which requires that she maintain an open-door policy one day a week. It also and crucially shows that even if society is vicious, virtuous people do not withdraw from it. The narrator's positive value is demonstrated by his refusal to engage in malicious gossip and fabrication, and by his distrust of improbable or shocking reports. But proper detachment does not include physical removal from the social scene according to *The Tea-Table*. Proper detachment is an ability to remain intellectually, emotionally, and morally uninfluenced by those who commit crimes against morality or sense. Similar to the people who go into company and behave in a selfish manner, the individual who leaves the company also behaves in a selfish manner. Both parties are uncivilized and both are to be censured.

Ultimately, Haywood's *Tea-Table* proposes that although neither the wholly sociable self nor the wholly isolated self can generate reliable knowledge, the self can generate reliable knowledge when it balances mental and emotional detachment with mental and social engagement. Knowledge with real use value, a key justifying motive for natural philosophy, is not achieved through isolated inquiry but through the combination of group endeavor and individual, disinterested acumen. Kathleen Lubey points out that Haywood advocates and uses this position in her amatory fiction, showing through her heroines the disastrous results of a failure to detach from oneself, and requiring of her readers the ability both to detach from the events of the narrative and to understand and be part of the community of feeling that narrative generates by creating an audience. "Readers must 'be sensible' of—that is, both aroused by *and* detached from—their own passionate 'falling' into the immoderate states of excess about which they read," Lubey argues. "Through this experience of absorption and detachment, Haywood's readers emulate the sexual pleasure that undoes her heroines, feeling the sensual force of the very temptations they must consciously monitor and resist in their own lives."[38] While Haywood's amatory fiction models the disaster that follows the failure of this combination of integration and detachment in amatory heroines, *The Tea-Table*'s narrator models its success. Both kinds of texts allow Haywood to advocate for a certain kind of knowledge and encourage her readers to attempt it through her narratives. Knowledge with value, particularly the moral value of improving the world, can only be achieved through the united strivings of critically independent, socialized selves. This is not the layered consciousness that Barker explores in the Galesia trilogy, discussed in chapter 3. It is instead a dynamic, a back-and-forth or oscillation, that provides in its resonance useful, deep insight.

It is also a dynamic more consistent with *The Spectator* itself than with the detached observation supposedly promoted by Mr. Spectator. *The Tea-Table, The Female Spectator,* and *The Spectator* amply testify that human nature as much as the rest of the natural world is well worth studying, but studying human nature requires a balance of sociability (one must go among people to do it) and detachment (one must think as if they are not there). When Haywood celebrates the value of "retreat" and "retirement" in *The Female Spectator* twenty years after *The Tea-Table*, she does not make it an absolute. The "Gentleman" who complains when his wife "with a gentle Force, dragg'd him from his Closet" does so not because he despises company but because sometimes he is not "disposed for Conversation" and prefers "to contemplate on the Blessings" she bestows on him.[39] Despite his claims of detachment, after all, Mr. Spectator is really, as Johnson might have said, a very clubbable man, and that clubbability is not simply a key character trait, but a *modus operandi* for him. The *Spectator*'s "Plan" is "laid and concerted (as all other Matters of Importance are) in a Club," he explains in his first issue. His parenthetical aside is worth considering, for with it Addison and Steele admit that whatever may be gleaned by isolation from others, "Matters of Importance" can only be pursued in a more collective or integrated setting. Furthermore, Mr. Spectator admits to being influenced by others. "I live in the World, rather as a Spectator of Mankind, than as one of the species," he claims, undermining this separation later by acknowledging that "my Friends have engaged me to stand in the Front."[40] By consorting with certain people, Mr. Spectator is as unreliable—or reliable—as Haywood's Amiana.

Initially, it would appear that what distinguishes the part of the narrative focusing on Amiana's quintet from the part describing the unpleasant visitors is just the people at the tea table. The narrator's first set of conversations are appalling, full of misinformation, disingenuity, selfishness, cruelty, and hypocrisy. The "Titled *Coxcomb*" is "resolved to Damn" a play because the playwright "had not consulted him in the Affair" (75). A young woman immediately "began to complain of a most terrible Headake, rolled her Eyes wildly round the Room, wreath'd her Neck, and distorted a Face which Nature had made extremely lovely" because she "fancies" that it "gives her an Air of Delicacy" that hides her conversational deficits—at least according to the "malicious" young woman who explains this behavior to the narrator (75–76). After those guests leave and the narrator returns to the tea table, however, the conversation becomes philosophical, edifying, insightful, and thoughtful. Instead of guests talking about each other, they talk to each other, and they agree and disagree in the most respectful terms. Certainly Haywood is making a point about the quality of group inquiry, or even the possibility of group inquiry, depending on the group's membership.

The narrator's presence at both conversations—that continuity of structure—is also crucial, however. His dispassionate reporting from a nonparticipant's perspective confers credibility on both accounts. The first set of conversations and the second, group conversation are equally reliable data, because the same reliable observer recounts each. However shocked and disgusted the narrator might be with the behavior of the first set of guests, and however pleased he might be with that of the second set, it is his own character and behavior that guarantees the soundness of the comparison and the conclusions that may be drawn from making it, as well as from each set of descriptions.

This use of the narrator is central to the defense of fiction and the terms on which that defense is conducted in the final third of Haywood's narrative. This section of *The Tea-Table* features "Beraldus and Celemena," the story-within-a-story, followed by an idealized discussion of the novella. In other words, Haywood models what novels can do in a debate within a novel about what novels can do. This part of Haywood's novel is a validation of the text in which it is embedded: anyone seeking to mock or discredit *The Tea-Table* as narrative fiction must first contend with Haywood's built-in justification of the text within it. But Haywood was always concerned with the role of fiction as a social force, and this section of *The Tea-Table* extends that conversation from the particular text to a defense of fiction.[41] Embedding an amatory novella within *The Tea-Table*, which is not amatory fiction, allows Haywood to highlight the conventions that not only define amatory fiction but also serve epistemological purposes. The conversation in the final third itself elucidates aspects of Haywood's defense, but the larger structure of *The Tea-Table*, a story recounted by a narrator that renders obvious the positioning of different aspects and episodes of the narrative, also contributes to the defense of fiction. In the final third of her narrative, the story-within-a-story of Beraldus and Celemena shows how narrative fiction is an authority different from other forms of authority, yet is also valuable to its society: according to Haywood's *Tea-Table*, narrative fiction plays a crucial role in transmitting knowledge and socializing others. It is a moral force in and of itself by virtue of being a form of witness. By claiming that amatory fiction provides moral instruction and reinforcement, Haywood also attempts to legitimize discursive space otherwise valued negatively as being both female and merely entertaining. Haywood's bid for a rational, co-ed environment and her claims for the value of narrative fiction are a way of resisting the silencing of women in the construction of a certain kind of self and a certain kind of knowing that goes along with that self. Furthermore and at the same time, by creating witnesses out of its readers, narrative fiction also creates individuals who are moral and socialized, as well.

The defense of fiction begins when Brilliante reads "A NOVEL" entitled "Beraldus and Celemena," to Amiana's guests. "Beraldus and Celemena" is preoccupied with the dangers of detachment in the sense of being removed by society. Like many an amatory protagonist, Celemena is a marginal female but unlike many of them, her lineage is tainted, underscoring her difficult position: she is the illegitimate daughter of a licentious, impecunious prince. Her father had "an Aversion to marriage" yet "had, by a Lady of no mean Rank, a Daughter," surely no testament to his virtue, sins compounded by an inability to provide for Celemena (86). To remedy the situation, she is sent as a lady-in-waiting to the court of the Princess of Parma, whose virtues are famous: "Never was a Woman fam'd for more Perfections than that excellent Princess" (87). Here however Celemena is also an outsider: younger than the rest, poorly educated in both life skills and intellectual pursuits, and a foreigner. The Princess, who loves her, considers Celemena a "an Orphan entirely committed to her Charge, having neither Parent, Relation, nor Friend near her, to whom she cou'd apply for Advice in any Affair, nor fly to for Protection if injur'd," in short, a "friendless Innocent" (90). Celemena is still more marginalized when she is seduced by Beraldus, a courtier, and she becomes common gossip (88–89). Finally, when Beraldus is punished for seducing her by being married off to a different—and enthusiastically immoral—woman (100), Celemena is pushed out of society: she does not have the opportunity to marry her seducer, which would have redeemed her reputation, reintegrated her into society, and made her happy. In punishing the criminal the prince and princess neglect to provide a happy ending for Celemena. She ends the narrative as far removed from society as possible: after attacking Beraldus' bride, she enters a convent and then dies (102–3). Ultimately, rather than offering her an advantage, being removed from others leads to her destruction.

Celemena's outsider status is exacerbated by her decisions. She is hardly responsible for her birth or the ill-will of the waiting women but the narrative also makes it clear that Celemena squanders an excellent opportunity to enter society fully. Her father's friend, who advises her father to send her to the Princess of Parma, "knew if Celemena…were once received into her Family, it must be wholly her own Fault if she were not in a State rather to be envy'd than pity'd" (87). Although Celemena is despised by the other waiting women, the Princess protects her out of affection (87–88), but Celemena ruins her opportunity for integration when she falls for Beraldus' wiles and then refuses to confide in the Princess. Celemena repeatedly rejects opportunities to confide in and therefore establish a connection with the Princess, which in turn would have preserved her position within society, however tenuous.

Haywood compounds Celemena's social detachment with a failure of intellectual and emotional detachment: any benefit that separation from the group might have offered is obliterated by her inability to think critically. As Haywood does with so many other characters, she makes Celemena believe in the appearance, not the reality, of the plausible lover. Celemena's failure to use the resources of her society, especially but not only the Princess of Parma, means that she fails to think analytically and therefore to act knowledgeably and virtuously. "The Confidence she had in [Beraldus] made her as little careful in concealing her own Passion, as she was in searching into the

Validity of his," the narrator explains (88). She is a "believing Maid," "the too Credulous CELEMENA," and "She, who always found invincible Reason in every thing he said" (88, 91, 93). The problem is not only Beraldus, although of course he's no help, but also Celemena. Because "her ador'd BERALDUS had said 'twas otherwise, and that was enough to make her assur'd of its being so" (94), she misinterprets the Princess's concern for her, burns the only evidence she has that she was seduced, and eventually temporarily runs mad: after trying to murder Beraldus' bride, "she remained a long Time wholly incapable of Reason" (102). Haywood's narrative offers a plausible enough tempter but it focuses on Celemena's failure to use reason, knowledgeable authorities, and critical thinking as the cause of her destruction. Celemena is not a good reader, in other words, a weakness introduced by being an outsider and then reinforced by isolation rather than remedied by it.

Haywood also calls attention to Celemena's failure of reason in the analysis of the narrative that follows. Amiana complains that Celemena is a badly drawn character because "She yields...with too much Ease, to create that Pity for her Misfortunes, which otherwise they cou'd not fail of exciting" (103), but Celemena's inability to generate pity is part of the purpose. As Amiana acknowledges, women should use their reason to counter the "Excess of Passion" that could be generated by "the agreeable Person of a Man.—If there were no Measures to be taken to secure ones self of his Affection, there are certainly to discover if he has Wit, Honour, and Good-nature" (103). Philetus' answer reinforces the interpretation that the problem is not love itself but the failure of the woman to analyze her situation without the influence of love or her senses. He reminds Amiana that "if the Ladies always made use of their Penetration, and chose for their Favourites only such as were worthy of them, there wou'd be no such thing as Woes in *Love*" (103). Similarly, when Amiana complains that reading about a villain like Beraldus is too shocking, "But yet sometimes 'tis necessary, said BRILLIANTE, to be reminded that there have been Men so base; our Sex is of it self so weak, especially when we suffer what little share of Reason we have to be debilitated by Passion, that we stand in need of all the Helps we can procure," a position reinforced by Dorinthus, who adds, "I am so far of your Mind, Madam...that these kind of Stories are of great use to persuade the Ladies to make use of that Penetration which AMIANA just now recommended" (104). The conversation raises questions about Celemena's behavior, then answers those questions by explaining how it is the failure of critical thinking, that is, thinking about the man and his actions separate from the feelings his person or actions might generate, that is the root of the tragedy.

"Beraldus and Celemena" advocates reason and the idea that women can be and ought to be rational. Celemena's behavior never appears inevitable because she is a woman. The Princess of Parma and the expectations of Celemena's father and his friend show that Celemena could have chosen to act differently had she been not only better educated but also more willing to do some intellectual heavy lifting. The discussion that follows this exposition, however, goes further, to assert the moral value of narrative fiction, a claim that Haywood makes elsewhere on behalf of and through amatory fiction, as Lubey points out.[42] Philetus says fictional narratives are not "design'd, as some imagine, for *Amusement* only, but for *Instruction* also, most of them containing Morals, which if well observed would be of no small Service to that that read 'em" (104–5). He adds that the illuminati of narrative fiction—"Madam D'ANOIS, Monsieurs BANDELL, SCUDERY, SEGRAIS, BONAVENTURE Des PERRIERS, and many other learned Writers"—would never have written for trifling reasons: "They had an Eye to the Humours of the Age they liv'd in, and knew that Morals, merely as Morals, wou'd obtain but slight Regard: to inspire Notions, therefore, which are necessary to reform the Manners, they found it most proper to cloath Instruction with Delight" (105). The terms of this defense are significant. Narrative fiction presents a general moral through particular instances, it reinforces the rational over the irrational, and it helps readers learn to think critically and in a detached manner about things near and dear to their emotional selves. These are valuable offerings.

Haywood further underscores the importance of thinking by the repeated use of the word "mind" in this discussion. Amiana's complaint about the narrative's "Shock to the Soul," is answered with a hail of references to the mind. "But yet sometimes 'tis necessary, Said BRILLIANTE, to be reminded," "I am so far of your Mind, Madam, answer'd DORINTHUS," "In my Mind, therefore, rejoyn'd PHILETUS," and Philetus shifts to an overt defense of novels with the phrase "methinks" (104–5). This emphasis on the mind and thinking is followed with the direct statement of the novel's legitimacy and function, its reinforcement of morals and virtue (105). It is not enough to feel. Haywood's defense of the novel says that readers' response to narrative must also be analytical—readers must think about it, not simply respond to it. The audience must be detached from the narrative even as the audience engages with it. That is how they can become knowledgeable from it. Furthermore, this third section does not stand alone. Structurally replicating the balance of separation and integration, the third section is connected to the rest of the narrative through a detached narrator who does for his own experience what the characters struggle to do with the fictional experience of Celemena.

Haywood introduces "Beraldus and Celemena" with Brilliante's statement that, "since we have insensibly fallen from the particular Passions, to the Temper in general, I will read you a Manuscript Novel which I put into my Pocket" (86). As Peter Dear points out in *Discipline and Experience*, the use of the particular to illustrate the general was a standard maneuver in natural philosophy well into the eighteenth century.[43] It certainly had been in literature for some time, such as in the use of exemplars. Haywood calls attention to "Beraldus and Celemena" as an example of the "general," not of the "particular." The plot, the characters, the conclusion—these are all to be used as instances of "the Temper in general," that is, of humanity. Already, then, Haywood makes a claim for her novel-within-a-novel: that despite the appearance of specificity, with particular character names and actions, the narrative reveals a general truth about humanity. It is a claim consistent with *The Tea-Table*'s opening statement that "*I have no View to any particular Persons, or Families in the Characters contained in the following Sheets.—The Design of them being only to expose those little Foibles which disgrace Humanity*" (73). About "Beraldus and Celemena," Brilliante adds that its virtues include "some Variety in the Shortness of it, and also an excellent Moral for the Subject," to reinforce its claims to ideas about the generality of humanity (86). Philetus announces that "We become virtuous ere we are aware, and by admiring the great Examples which in the narrative appear so amiable, are led to an Endeavour of imitating them" (105). In other words, narratives about particular people, even fictional ones, are capable of carrying general truths.

Truths are extracted by the group discussion that follows. Significantly, Amiana is not the character, or even *a* character, who articulates the position ultimately favored by Haywood, as she is the one who first raises objections to the amatory fiction-within-a-fiction of Beraldus and Celemena, and then later claims that "if I were of Counsel with the Writers of such Books, I shou'd advise 'em to chuse only such" that explore the "many Misfortunes to be found in Love, even where both Parties are perfectly sincere" (103, 104). This last claim is fairly ironic considering that she is the creation of one of the "Writers of such Books" and helps refute the view that Amiana is Haywood's persona.[44] Furthermore, consistent with the other discussions at Amiana's model tea table, no one character in this analysis of "Beraldus and Celemena" provides a complete or completely convincing position. Only the combined arguments of Philetus, Dorillus, and Brilliante can fully overcome Amiana's objections (103–5). In this scene, Haywood models the use of amicable conflict as a tool for uncovering truth, showing a collection of independent thinkers working together to tease out and articulate authoritative insight.

At the same time, this conversation itself is recounted by a narrator who offers it to his audience. If "Beraldus and Celemena" generates an audience separated from its events—Amiana and her guests—then the story of the auditors receiving and discussing "Beraldus and Celemena" is generated in turn by an audience separated from *its* events—the unnamed narrator, who does not contribute but does report. Haywood's narrator provides a detached account of events to which he is, nevertheless, connected. He is connected through sympathy with the players such as Amiana and Dorinthus. He is also connected through his modest witnessing, which presents the scene for its use-value to his own audience. The narrator is outside the narrative he presents and yet imbricated in it, just as the audience of "Beraldus and Celemena" is outside the narrative but invested in it.

As Kathryn R. King points out, meta-fiction is not a mode that is usually associated with Eliza Haywood's narrative fiction.[45] King suggests that in *A Spy upon the Conjurer* (1724), however, Haywood satirizes the kind of reader that her work appeals to and creates. In Justicia, Haywood creates a narrator whose rapacity for narrative titillation drives her to increasingly transgressive acts, representing the reader of amatory fiction who is always in pursuit of information and sensation.[46] Leah Orr questions the attribution of *A Spy upon the Conjurer* to Haywood but the similarities to *The Tea-Table*, written so close in time, are suggestive even if inconclusive.[47] If *A Spy upon the Conjurer* is Haywood's, then she wrote more than one work of meta-fiction, which would be consistent with the repetition of theme and technique in her *oeuvre*. Furthermore, the works can be seen as companion pieces: Haywood would have indicted unregulated, uncritical, isolated reading in Justicia, the narrator of *A Spy*, and advocated interactive, critical, contextualized reading in the narrator of *The Tea-Table*. Like so many aspects of Haywood's work, the similarities are tempting, however inconclusive.

Regardless of the status of *A Spy upon the Conjurer*, the narrator's position in *The Tea-Table* is significantly different from that of the narrator of *Part 2*, which was definitely written by Haywood. In *Part 2*, the narrator admits to the reader that the poem to Hillarius that opens *Part 2* is his own. After Amiana reads the poem, the narrator confesses that "I, who had a greater Interest in the Author's Fame than any Person in the Company excepting Amiana, was sensible of, had the Pleasure of observing something in their Countenances, which inform'd me that the Respect they paid to that Lady was not the only Motive of their Approbation," thus acknowledging his emotional investment and literary participation. The narrator also embodies himself by mentioning his "Pleasure" and careful observation.[48]

He is not the only character in *Part 2* with a body. Haywood also acknowledges a subtle chemistry among Amiana's circle that requires a body and the feelings catalyzed by one. Early in *Part 2*, the narrator says that no one would have criticized the poem that Amiana read, but then notes that "Philetus was so far from condemning, that taking the Paper out of her Hand, and telling her in a gallant manner that to judge impartially of any thing, it must not appear under the advantages of such a Voice and manner as her's who read it, he look'd it over several times, and at his returning it, assur'd her with an Air of Sincerity" that it was very good (44). The passage notes the roles of a hand and a voice, uses "manner" twice, and concludes with Philetus' "Air of Sincerity" that raises the specter of deceptive appearance. Philetus' point is that the body can influence aesthetic judgment. Later, the characters disagree about the role of the body in marriage (51–54). Amiana rejects a poet's conclusion that "Men's Hearts are not to blame one bit" for marital discord because "Our Souls would never disagree, / If once our Bodies did but fit" (52), a rather suggestive ending in such a seemingly chaste company. "To Hillarius," the poem that opens *Part 2,* provides another suggestive moment, when Alexis says to Clorinda of Hillarius, "His Beauty thy admiring Eyes have seen, / His Courtly Air, his soft engaging Mein, / His flowing Wit thy list'ning Ears has blest, / And as thou may'st, thou guessest at the rest" (40). *Part 2* appears to be more sensitive to and more willing to acknowledge the sexual dynamics of any mixed group; one might speculate that in this way Haywood acknowledges the chemistry that disrupted the Hillarian circle, as well. Certainly the recurring theme in *Part 2* of love gone spectacularly awry suggests a more personal resonance than the more philosophical debates about the definition of love and the passions in *The Tea-Table*. The possibilities of autobiographical resonance in *Part 2* notwithstanding, however, there is a significant difference in the handling of the body and a group of bodies between the original *Tea-Table* and its sequel.

In fact, the people of the first *Tea-Table* are surprisingly disembodied. "Surprisingly" because this is a Haywood novel and she knew a thing or two about how bodies can and do operate, and because it is a narrative exploring the contextualization of knowing by an author highly sensitive to the role of gender and the body in providing access to knowledge. What could explain the bodilessness of the narrator in *The Tea-Table*, for example, a bodilessness so profound that scholars disagree about whether the narrator is a man or a woman? Alexander Pettit, Margaret Case Croskery, and Anna Patchias assume that the narrator is a woman, for example, a gender that Haywood usually chooses for her narrators.[49] The few clues to the narrator's identity and the narrator's behavior during the novel point instead to a male narrator, however. He has been introduced to the group by Dorinthus, obviously a man, and it would have been very unusual for a male to introduce a female friend to a tea table without further contextualization. Dorinthus also presumes to chastise publicly the narrator for rudeness, unusual behavior for a man toward a woman in a public setting but less so for a man toward another man. The narrator also is familiar with the theater scene that the Coxcomb draws on but is astonished by the behavior of the hypochodriac and her false friend, an uneven familiarity suggesting male experience. While the narrator's naïveté is an old, useful device for defamiliarizing—particularly in travel narratives and utopia narratives, for example—in this case, the naïveté is selective not comprehensive, suggesting unfamiliarity with certain gendered realms rather than with everything. Even with these clues, however, there is little in the depiction of the narrator himself to answer the question definitively. As a source of knowledge, he is clearly embodied—his presence at the tea table is crucial to his functioning. After all, as Pettit notes, this is a work depending on location.[50] But the function or participation of that body is minimal.

The same can be said of the other characters in *The Tea-Table*. While in *Part 2*, as I have noted, Haywood infuses the interactions of the characters with an erotic charge, this charge is essentially absent from the conversation of *The Tea-Table*.[51] The characters are described primarily in terms of character, rather than person. Philetus is "A Gentleman than whom there is scarce to be found one Master of more Accomplishments," Brilliante is both "Lovely," whatever that means, and "Witty," and Dorinthus "has few Equals for fine Sense" (77). Amiana is a lady of "Quality and Merits," "agreeable," and full of "Sweetness of Disposition" that inclines her to "Magnifie" her guests' virtues (76, 74). They are present and the fact of their being present is crucial to the purpose, but what body is present is less crucial. Haywood seems to be following the emerging and Lockean emphasis on the disembodiment of knowledge, the idea that the body is not only disconnected from knowledge, but also disconnected from knowledge production. As Steven Shapin points out in *Never Pure*, during this period "the truth-seeker is someone who attains truth by denying the demands of the stomach and, more generally, of the body."[52] In *The Tea-Table*, Haywood certainly rejects the idea that the body inevitably determines knowledge or the self's ability to know or to produce knowledge, by rendering all her characters' bodies the same—absent—in the presentation of those characters as producers of knowledge. At the same time, however, because all of Haywood's knowers must have a body, she is not rejecting the idea that the whole self, body and mind combined, has a role in the creation, reception, and transmission of knowledge. Rather Haywood rejects the body itself as the sole determinant, and with it the idea that the gendered body distinguishes those who can discover, communicate, and process knowledge, that is, men from those who cannot, that is, women. The problem is not embodiment; like with the thoughtful mind, it is a question of what one does with the body that shapes how the self interacts with knowledge.

As Haywood demonstrates, in fact, unregulated bodies are distinctly troublesome. The Coxcomb is introduced with the physical "came into the room." The hypochondriac uses her body to command attention (75). Her behavior is explained by a woman with "so malicious a Look and Accent" that she drives the narrator to a window seat (76). At the end, the polite and edifying conversation is interrupted by a "Lady in a new Suit of Cloaths," who calls attention to her body by demanding everyone talk about her attire and calls attention to other peoples' embodiment by discoursing on their wardrobes. She complains that "Lady BELLAIR discovered an unbecoming Assurance in exposing her bare Neck, because it was the only handsome thing about her. Lady PRUDENCE, to conceal the Deformity of hers, sweated in *July* under the Weight of a Scarf and half-a-dozen Handkerchiefs" (106). All these characters parade their bodies in order to attract attention, particularly from the opposite sex. The Lady in a New Suit's monologue has its own racy undertones, focusing on dress and the body beneath (one woman she complains of "had her Petticoats too scanty," the narrator reports, seemingly oblivious to the innuendo).[53] In this way the negative characters who bookend the conversation are too invested in their bodies, their selves too much composed of their bodies, to be reliable sources of information or credible partners in the search for truth. Like Haywood's tragic amatory heroine Celemena or any of her other tragic amatory heroines, the unchecked acting of their bodies renders both men and women irrational and therefore unreliable. Drury suggests that the discussion of the four conversants on the role of the body and passion allows Haywood to articulate a Hobbesian mechanism and to accept the double standard, but Haywood's analysis of the role of the body goes beyond her use of Amiana's favored guests.[54] All the negative characters are too corporeal, just as the narrator threatens to be too cerebral when he removes himself from company. Moral authority does not lie in a body, in a mind, or in the removed individual. Society, in other words, has a role just like the individual consciousnesses do.

For some authors, the ideas about the self that had been introduced during the late seventeenth century provoked questions about the value and function of the knowledge that it produced as a result of its own position. That position, as Haywood pointed out, had a great deal to do with its gender because of the way gendered bodies were valued and placed in early eighteenth-century England. Her claim that the self can only achieve authoritative, reliable, and moral knowing and knowledge when it is able to think independently but operate connectedly, implicitly rejects the relegation of women, their bodies, their experience, and the knowledge that comes from the combination of those forces, to the margins. In *The Tea-Table*, Haywood uses narrative structure to propose an alternate position for the self, to enfranchise socialization and dialogue. While she may have lost the battle when it came to the valorization of individualism and the notions of property, law, and government that gave rise to crucial legal and political events of the later eighteenth century,[55] the idea of a knowledgeable but detached raconteur has certainly stayed with the novel. The detached observer is a significant component of literature's omniscient point of view, itself a key convention in the novel's cornucopia of attributes. In this regard, Haywood's model in *The Tea-Table* is neither the last nor the only word on the question of the morality of knowledge and the new self, nor on the role of point of view in establishing the novel as a genre. As chapter 5 argues, Mary Davys, Haywood's contemporary, had a different and more optimistic view of the relationship between detachment and morality.

Chapter Five

The Moral Observer

Mary Davys is the author in this study who is the exception that proves the rule. While Aphra Behn, Jane Barker, and Eliza Haywood all had connections to natural philosophy one way or another, Mary Davys did not. She knew Cambridge students, but there is no evidence that her acquaintances were or went on to be natural philosophers. Nor is there evidence that Davys read works of natural philosophy, although her novels suggest that she was familiar with natural philosophy from its presence in popular culture. It is precisely because of the nature of this connection to natural philosophy that Mary Davys's novels concern the final chapter of this book. Davys's work reveals an ongoing interest in the relationship between knowledge and morality, detachment and integration that occupied so much of natural philosophy during this period. Like Eliza Haywood's *The Tea-Table* (1725) discussed in the previous chapter, Davys's novels investigate how a self can be not only reliably truthful because disengaged but also reliably judgmental because morally correct. Unlike Haywood's narrative, which explores the issue in terms of the social position of the narrator, Davys's narratives consistently consider the issue in terms of the nature of the self. In attempting to create a self who is internally morally invested but narratively detached, Davys draws on contemporary, if debated, ideas of the self and develops techniques that underpin literary omniscience.

Mary Davys's novels are products of an epistemological moment in which the values of accuracy and detachment and the values of morality and community were being negotiated. As I also discuss in chapter 4, natural philosophy's premium on dispassionate observation and analysis was sometimes reconcilable with and sometimes in conflict with ideas about the use-value of knowledge and the role of morality in generating and confirming knowledge. Although there is no evidence that Davys read John Locke or Isaac Newton, Davys's novels, like Haywood's *Tea-Table*, are structured to respond to and engage with these ideas. But where Haywood's interest in detachment, knowledge, and morality has to do with social integration and separation, Davys's interest has to do with the individual him- or herself.

These concerns appear in other aspects of Davys's novels, particularly her themes, as well, but Davys's use of narrative structure to investigate the relationship between detachment and inclusion, knowledge and morality, characterizes her work over the nearly three decades in which she wrote and provided techniques that later novelists could and did use.

Because they were inflected by an epistemology that valued accuracy and detachment on one side and morality and usefulness on the other, Davys's novels are often classified as either realist or didactic, classifications generally presented as mutually incompatible.[1] As a realist author, Davys is grouped with men as if women did not write in a realist mode; as a didactic author, she is grouped with women as if men had no interest in didacticism.[2] Such monolithic categories cannot hold: there are female realist novelists and male didactic novelists.[3] A deeper problem with these classification schemes is that during this period novelists and philosophers, including natural philosophers, considered an interest in nature, including human nature and the psyche, to be entirely compatible with an interest in morality. "Moral knowledge," John Locke declared in his *Essay on Human Understanding*, "is as capable of real certainty, as mathematics."[4] In such an environment satire, realism, and didacticism can be mutually compatible; narrative innovation in the early novel can be driven by the interest in merging what Frans De Bruyn calls the "didactic impulse with a newfound psychological verisimilitude."[5]

These different classifications for Davys's work therefore can be reconciled when they are read in this epistemological environment. Davys's novels, particularly *The Reform'd Coquet* and *The Accomplish'd Rake*, consider the importance of the relationship between the body and the mind in constituting a self whose knowledge is moral and whose morality is knowledgeable, issues of considerable interest to natural philosophers and discussed at length by Locke. Davys's narrators serve to investigate the roles of detachment and engagement, and the roles of the body and of the mind. They thus respond to ideas about knowledge and the body influentially articulated by Locke in his *Essay*. "*Self* is that conscious thinking thing, (whatever substance made up of whether spiritual, or material, simple, or compounded, it matters not) which is sensible, or conscious of pleasure and pain, capable of happiness or misery, and so is concerned for *itself*, as far as that consciousness extends," he explains (II. xxvii.17). The body is "vitally united to this same thinking conscious self," but the self does not depend on the body: "Self," Locke states, "depends on consciousness" (II.xxvii.11; II.xxvii.17). Any body will do, but not any consciousness will do in constituting a particular self, which means that although the body contributes to self, it and the mind, or the consciousness, can be detached. This detachability of the "thinking conscious self" from the body underpins Davys's narrative strategies: Davys's narrators are credible because embodied, but accurate because their consciousness is detachable from that body.

Davys's novels posit a convergence of knowledge and morality in the position of the detached, observing outsider. "Like male writers, such as Daniel Defoe and William Congreve, Davys confronts the problem of combining fiction with moral probability and therefore with 'truth,'" Sarah Prescott observes.[6] Davys was a deliberate craftswoman; as critics since William McBurney have acknowledged, she advanced a self-conscious, thoughtful theory of the novel.[7] In her preface to the first volume of *The Works of Mrs. Davys* (1725), Davys declared directly and through the repetition of words such as "probable" and "likely" her determination to create a narrative that was feasible, entertaining, and morally satisfying: "*The Adventures, as far as I could order them, are wonderful and probable,*" she announces, "*and I have with the utmost Justice rewarded Virtue, and punish'd Vice.*"[8] To support this endeavor throughout her fiction, Davys uses structure to model complete and moral knowing: her narrator provides the moral framework as well as the factual framework within which all action takes place. Davys's narrators are removed from the action but very much part of the moral framework of the story, connecting to the characters structurally rather than personally. She is no nebulous, all-seeing rapporteur; this is the 1720s, after all, and character and wit matter. But in her ability to know everything including the values of the story and to tell it in a distinctive, seemingly existing way, Davys's narrators present an image of the ideal natural philosopher, if not the ideal self, at work.

A willingness to innovate narratively to pursue this interest in the role of detachment—individual from group, mind from body—characterizes Davys's work from the beginning of her career. In the dedication to one of her earliest novels, *The Fugitive* (1705), she uses "experiment" to describe her work: "I had a Mind to make an Experiment, whether it was not possible to divert the Town with real Events, just as they happen'd, without running into Romance."[9] This attempt at something new, at a new kind of narrative to support a new epistemology, appears in her structural choices, as well. The narrator or perspective from which the story is told is often removed in some way, such as the Irish widow of *The Fugitive* and *The Merry Wanderer* (1725), or the narrator in *The Reform'd Coquet* (1724) and *The Accomplish'd Rake.* In other narratives, the structure high-lights the moral framework in which the narrative operates, such as the frame narrative of *The Lady's Tale* (1725) or the epistolarity of *Familiar Letters* (1725). What interests Davys is the presentation of events within a moral framework: full knowing and full moral know-ing without the burden of engagement.

It is tempting to consider the personal appeal of such a view-point for Davys. There are few established facts about Davys's life. We can be certain that she was married in Ireland to Peter Davys, that he died young and that she also lost at least two daughters as small children, and that shortly thereafter she moved to London, then to York, then back to London, and finally to Cambridge. She cor-responded with Jonathan Swift, who told Esther Johnson ("Stella") that Davys begged for money, and who had known her late husband in school. In addition to Swift, Davys seems to have been connected to other Scriblerians—John Gay subscribed to *The Reform'd Coquet*, Alexander Pope subscribed to *The Reform'd Coquet* and *Works.*[10] She wrote novels during both stays in London, at least one play in York, and poetry in all three places; she operated a coffeehouse in Cambridge from at least 1718; and she died impoverished. From these facts we can glean that like Jane Barker, her contemporary, she had several marks of the outsider: she was a widow almost her entire life, she was associated with Ireland, and she did not have other kinds of status to compensate for these strikes against her.[11] It is no wonder if she was interested over the course of her writing career in the value of the outsider's perspective, in the idea that being an outsider offered a special value. It is dangerous business to ascribe biographical rea-sons for literary performances, however; the temptation to reduce creativity and intellect to experience is great, and the quality of the insights derived therefrom inconsistently reliable. In Davys's case, it is particularly necessary to respect but not depend upon biography and chronology since establishing the facts of her life is currently nearly impossible. Certainly, there is a provocative connection between the facts of her life as we know them and the recurring fascination with the position of the valuable outsider as expressed in her themes and even more consistently, in her narrative structure. That fascination has much to suggest about the evolution of the novel and its context.

Davys's novels evince an interest in the relationship between exclu-sion and inclusion that roiled epistemological waters during the late seventeenth and early eighteenth centuries. Like Haywood and many philosophers discussed in the previous chapter, Davys struggled to reconcile the advantages offered by critical and even literal distance from events, characters, and ideas with the value of morality which, by implicit definition, required adherence to and identification with a group. The discussion of Davys's work begins with *Familiar Letters* and *The Cousins* (1725), both published in *The Works of Mrs. Davys* but probably written earlier, to consider Davys's interest in narrative structure in forming patterns of knowing.[12] This section includes a consideration of *The False Friend* (1732), a republication of *The Cousins*. These novels can function as keys to the code of Davys's concerns and use of narrative structure in executing those interests. In the second section of this chapter, I discuss two pairs of narratives radiating out of this hub of works. Davys's first novel, *Alcippus and Lucippe* (1704), its reissue in the *Works* as *The Lady's Tale*, *The Fugitive*, and its reincarnation in the *Works* as *The Merry Wanderer*, all show ways in which Davys explored strategies for creating a detached, moral point of view for telling and understanding the story. This chapter ends with *The Reform'd Coquet* and *The Accomplish'd Rake* to show how Davys's most supple and complex work establishes a balance of detachment and morality and provides a foundation for literary omniscience. Taken all together, these novels reveal that Davys was engaged in a long-term investigation of the knowing self that intensified in the 1720s, the same decade in which Jane Barker and Eliza Haywood were also using narrative structure to explore the same issue, also with genre-building implications. The relation between separation and integration was central to Davys's narrative innovations and her contribution to the development of the novel as a genre.[13]

Familiar Letters and *The Cousins* are significant for what they reveal about Davys's interest in structural experimentation and her reasons for that experimentation. If Jane Spencer and Martha Bowden are accurate about their dates of composition, we can conclude that Davys was experimenting with narrative form throughout the twenty-five years of her writing life. If we rely solely on their publication date, the evidence still suggests that Davys was innovating in the middle of her second burst of publishing, as they are structurally inconsistent with her most sophisticated and successful novels, *The Reform'd Coquet* and *The Accomplish'd Rake*. Both *Familiar Letters* and *The Cousins* indicate that Davys was using narrative structure to balance the connection among detachment, knowledge, and morality.

Davys's only epistolary novel, *Familiar Letters* often attracts attention for its courtship narrative and for its declaration of Whig principles. Toni Bowers and Alice Wakely have called attention to epistolary "seduction" narratives as political instruments and both directly and indirectly, *Familiar Letters* can be read within a Whig context.[14] Spencer notes that the political argument is the means by which the characters express their romantic anxieties about their relationship, and also that "Their stories encode their politically based anxieties about becoming involved with each other."[15] In this regard, however, it is important to understand "Whig" in more than strictly political terms. Lockean epistemology underpinned Whig politics, the developing British state, and the inextricably entwined simultaneous developments of natural philosophy. Susan Glover argues that Davys's Whig politics appears in her handling of property in *Familiar Letters* and a Lockean sense of property appears in the prefatory materials to *The Reform'd Coquet*.[16] Davys's Whiggishness is more complex than support for George I. It is expressed also in her views of knowledge and the self, as *Familiar Letters* makes clear.

Of all of Davys' novels, *Familiar Letters* is the one with the most connection to contemporary philosophical developments and discourse. There is, of course, the well-acknowledged reference to John Flamsteed, but the full passage suggests more at stake than a mere reference. Artander writes to Berina of

> the threescore Witches Mr. *Flamsteed* has found out in *Westminster*; I hear he intends to beg 'em of the King, and roast them by the Blazing-Star next *April*. I am just going to sink a Vault for a Retreat from its sulphurous Effects, and wou'd have you come and share the Advantage of it before the Conflagration begins. Those Men of Profundity in the Occult Sciences, divert themselves with other Peoples Fears, and laugh to see the intimidated World shock'd with Horror at their Prognostications. Last Winter, when I was at *London*, I remember the Lords of the Admiralty were very busy with a Gentleman who had for certain found out the Longitude; but having heard nothing of it since, I doubt the Project is dropp'd, and our Ships must fail as formerly.[17]

As Bowden points out, the reference to Flamsteed's "threescore witches" that he "beg'd" of the King with the intent to "roast" them in April could be a reference to Flamsteed's *Historia coelestis*, which the Royal Society under the direction of Isaac Newton pirated in 1712. Flamsteed got 300 copies of the pirated version back in March 1716 and burned all but the correct parts immediately following. Given that the copies arrived on March 28 and he first excised the section of each copy that was salvageable, it is likely that he did not start burning the mutilated pirated books until April.[18]

The Flamsteed reference is only one of many invocations of natural philosophy in this passage, however. The term "blazing-star" refers to a comet and seems to allude to a specific event, although there was no comet spotted in England between 1707 and 1717 (furthermore, almost no one noticed the one in 1717).[19] While the event cannot be identified, Davys is clearly drawing on celestial events within a discourse of natural philosophy. Davys also alludes to the longitude problem when Artander writes, "Last Winter, when I was at *London*, I remember the Lords of the Admiralty were very busy with a Gentleman who had for certain found out the Longitude." In 1714, Whiston and Humphry Ditton proposed a scheme for establishing the longitude, *A New Method for Discovering the Longitude Both at Sea and Land*, published again the following year.[20] Cannily pitched to "the Publick," Whiston and Ditton's slender work attracted popular attention and compelled the Admiralty to pay attention, if only to dismiss it. In response to public pressure, the government offered an enormous bounty in 1714 to the first person to solve the longitude problem and proposals began to arrive immediately.[21] Part of the drive to establish the Royal Observatory (and Flamsteed in it, as the first Astronomer Royal) was to create a celestial chart sufficiently accurate to enable sailors to compute longitude and sail safely, one possible explanation for the association in Davys's novel of Flamsteed, the "witches," and the longitude problem in this passage.[22]

Both the "blazing star" and the longitude references address popular controversies involving natural philosophy, an engagement with natural philosophy reinforced in Artander's contempt for "Those Men of Profundity in the Occult Sciences, [who] divert themselves with other Peoples Fears, and laugh to see the intimidated World shock'd with Horror at their Prognostications." Astrologers had long predicted disasters, and strange celestial events had long been associated with human and natural cataclysms both current and imminent. Furthermore, astrology and astronomy had considerable overlap during this period.[23] Michael Hunter points out that late seventeenth- and early eighteenth-century astrologers often possessed compendious and accurate observations of the skies, sometimes to the envy of astronomers.[24] The proto-scientific and popular interest in comets sparked by Edmund Halley's comet was readily converted into millenarian anxieties between 1714 and 1716.[25] The author of the apocalyptic *The Black Day* (1715), for example, announced that he used "the Corrected [tables] of that famous Astronomer Mr. Flamsteed" as the foundation for his predictions.[26] In Davys's formulation, "Men of Profundity in the Occult Sciences" have a pernicious, social effect. Having provoked frightened people into building bomb shelters, they "divert themselves with other Peoples Fears, and laugh to see the intimidated World shock'd with Horror at their Prognostications." Here Davys cannot be referring to Flamsteed, who did not publish "Prognostications." The allusion to Flamsteed and "threescore witches" therefore is perhaps not just a reference to the burning of his corrupt books but also a rejection of "Those Men of Profundity in the Occult Sciences." "Witch," according to the Oxford English Dictionary, could be used for a man or a woman during this time. Davys uses it to refer to a man several times in her *oeuvre* and always negatively.[27] These allusions suggest that Davys was familiar with natural philosophy's public manifestations and controversies. They also suggest that she had formed her own opinions about them before moving to Cambridge, refuting the theory that her knowledge and critical insight came courtesy of her Cambridge clientele.[28] It was very possible for her to acquire this knowledge on her own, especially in London, where publications, public lectures, and the like made such information readily available to men and women.[29]

These are not simply dropped contextual clues; natural philosophy and political ideology serve similar purposes in *Familiar Letters*. Like Jane Spencer, Alice Wakely views the political debate between Artander and Berina as an unsubtle code for the correspondents' debates over marriage.[30] In fact, Davys uses a collection of discourses to provide a framework for talking about, encoding, and decoding Artander and Berina's relationship—where it is and where it is going—that includes natural philosophy. Natural philosophy appears throughout the *Familiar Letters* and is always associated with Artander. As with politics and literature, Artander's references to natural philosophy are, and are more than, deployments of discourses for framing the dynamic between Artander and Berina.

Natural philosophy becomes a primary discourse when Artander attempts to introduce philosophy to "divert" Berina from politics. Davys is here punning on "divert," using the sense of entertain but also of derail. Artander writes:

> …you will comply, when I beg of you to put a stop to this sort of Correspondence; and let your Letters for the future, be fill'd with the innocent Diversions of the Town: 'tis a pity *Berina*'s Temper should be ruffl'd with Politicks. I am now going to divert you with something of a different kind. Yesterday being in a very philosophick Strain, I was resolv'd to visit Nature in its most private Recesses, and enter the Hollow of an adjacent Rock, of which you have often heard me speak. (*FL*, 101)

Artander's "begging" Berina to stop talking about politics for her own sake because it is clearly making her upset smacks of the technique of silencing someone with the claim of doing it for their own good. That he attempts to substitute "the innocent Diversions of the Town" reinforces his efforts to "dumb down" Berina's interests and discourse.

Furthermore, natural philosophy as represented in Artander's writing does not have the heft or legitimacy that politics has in the earlier exchanges, suggesting that Artander is substituting an inferior discourse. When he describes his spelunking, he concludes that "Curiosity only makes the Vertuoso; and if I go on a little longer, I shall grow a perfect Sir *Nicholas Gimcrack*," comparing himself to the ridiculous title character of *The Virtuoso*, Thomas Shadwell's 1677 satire of natural philosophy and the Royal Society (*FL*, 101). Returning home from a meeting with a lovelorn friend, Artander "met with a Matter of some Speculation; for casting my Eye towards the top of a Tree, from whence I heard a rustling Noise, I saw a Crow, with a living Fish in his Bill: Pray send the Phaenomenon to the *Royal-Society*, for, methinks I wou'd fain know how those Creatures, (neither of an amphibious Breed) came together"; he is evidently unable to recognize that the crow caught the fish (*FL*, 106). Associating natural philosophy with Artander in this way reinforces the implication in the later passage that "men of Profundity in the Occult Sciences" use legitimate discoveries and methods to disguise their own insufficiency. This association highlights how natural philosophy can be used to cover or disguise the intellectual insufficiency of people unable to handle mysteries of nature such as longitude or why nice young ladies do not wish to marry the young men who woo them.

Artander's references encode the shifting power dynamic between the correspondents. Each use of natural philosophy by Artander is not simply a commentary on their relationship but also an attempt to draw the subject to romance and to gain the upper hand. When he insists that she stop writing about politics, he substitutes it with the story of a male penetrating a dark "Hollow" to discover a vulnerable woman (101–2), a metaphoric rape fantasy.[31] Furthermore, describing the half-starved, abandoned, nearly incoherent woman, Artander states that she "seems to have a better Notion of Virtue than cou'd be expected from her Education," a phrase suggesting that Artander has tried that virtue for himself after probing her cave (*FL*, 102). Artander's story about exposing the secrets of female nature through an uninvited penetration retorts upon Berina's political images of a state requiring the agreement of its people and Berina's rhetorical efforts to compel Artander to accept the King.

His other significant references do the same work. After Berina uses poetry to warn him off wooing—"*You see*, Artander, *what the*

Gamester wins: / From the first Hour of Play, his Woe begins" (*FL*, 104)—he uses poetry to celebrate and conjure romance, followed by his observation of the crow with the fish in its mouth. Here the coming together of two very different species—crow and fish, man and woman, enamored and unrequiting, Tory and Whig—is only seemingly inexplicable.[32] The "Royal-Society" will be able to explain it and in the process, show how that union is natural (*FL*, 106). In fact, the two species do not come together narratively, because Berina does not answer this letter. It turns out that she is having a very good time at someone else's house (*FL*, 108–9), so not only are they separated intellectually, emotionally, and politically, but also she is not even thinking of him during the time he is waiting for her letter. It is after she reminds him of her ability to be charmed by others that he yields slightly, returning to politics briefly to accept a modestly Whiggish position and using controversies in natural philosophy to acknowledge the potential falseness of that discourse. Similarly, when he describes a beautiful woman at "an Entertainment" (who is the beloved of his lovelorn friend and should not be the subject of his desiring gaze), Artander says that "the Eyes and Admiration of the whole Company [were] drawn by a magnetick force upon" her as if she, not he, has control over his body (*FL*, 112). He uses the idea that he is helpless in the face of this "magnetick" force to exculpate himself from ogling his friend's beloved ("I had considered her whole Person," he admits) and needling Berina into jealousy. His last use of natural philosophy—"magnetick force"—turns the tables on her, reminding her that he too is capable of being "diverted" by attractive, entertaining company; she in turn capitulates with her dream of a blinded Artander rendering her blind (*FL*, 112–113). Overall, allusions like the political debates and the use of romance reveal that Davys was well-informed about issues that concerned people of the day, but not that she was a philosopher. As such, the use of natural philosophy confirms my point that by the 1720s ideas that had originated in the philosophical revolution had entered the mainstream sufficiently to be taken up by people who were neither philosophers nor interested primarily in natural philosophy per se.

If natural philosophy is not Davys's primary tool in *Familiar Letters*, what is? Davys's primary technique for representing the intersection of complete, authoritative knowledge and appropriate morality is structure. Davys depends upon the interaction of the letters to present the positions of the characters and how and why they change. Epistolary form shows these interactions from a third-party, entirely detached position. The letters comment on each other, often in pairs but sometimes in sequence. The interaction among the letters outlines the conflicts and their progress. When Artander writes Berina for the first time, complaining of his separation from her and fearing that she will not keep her promise to be his friend and correspondent, he tells a story of a young woman who treats her lover with contempt and is publicly shamed (*FL*, 93–94). Whether Artander is aware that he is telling this story as a warning to Berina not to treat him badly, Davys certainly is, ending the letter, "How happy are you and I, who have made the strongest Resolves against the Follies of Love! Be sure, *Berina*, keep your Friendship inviolate, and you shall find I will keep my Promise, in never desiring more" (*FL*, 94). Berina replies by affirming her friendship and then getting her revenge for his threats. She celebrates Whig politics and the Hanoverians, insults him by calling his political beliefs "poison'd Principles" and "Errors" but disingenuously claims that she is not insulting him, and concludes by congratulating the English on their "Liberty" and "Freedom" (*FL*, 95–96). It is surely no accident that Davys dates the letter in which Berina asserts Whig principles, the value of freedom, and the rejection of "a Yoke that galls for Life" to November 5, Guy Fawkes Day (*FL*, 96). In this way the letters answer each other, reinterpreting the code and the message. Each letter addresses a fear or hope in their relationship, using different strategies to express and persuade, and the next letter (sometimes a series of letters) responds in a similar manner. Each epistle offers a snapshot of a moment in the emotional life of its author; together, the letters are a film of the development of each character and of their relationship. In so doing, Davys exposes the problems in the Artander-Berina relationship.

The subject matter and rhetoric also balance or oppose each other. Davys signals Artander's ridiculousness with the sheer number of physical accidents that befall him during their exchange on politics. He trips over a woman's dress and they fall down the stairs, and he slips on a piece of old bacon and winds up in an old lady's lap (*FL*, 97, 99). Berina, on the other hand, is not only in command of herself physically but also in command of the rhetoric of the time. Despite his efforts to trivialize their subject matter and diffuse the tension, Berina insists on replying with powerful rhetoric to each of his comic scenes with over-powerful women, at one point describing "Barbarities" in graphic language—"the Clergy's Mouths cut from Ear to Ear, their Tongues pull'd out and thrown to the Dogs...Mens Guts pull'd out and ty'd to each other's Waists, then whipp'd different ways...Children ripp'd out of their Mother's Womb, and thrown to the Dogs" (*FL*, 97). Similarly, after Artander tells the story of the girl in the cave, Berina replies by telling a story of a man in a house so rudimentary and dark it might as well be a cave. This time, the creature lurking in the darkness is male, corrupt, diseased, and vice-ridden rather than female, innocent, and almost untouched by civilization; being "half-starv'd" in Berina's tale is not romantic as it is in Artander's, but disgusting and agonizing; and the person who probes the darkness into the inner sanctum is alone, frightened, abused, and forced to flee in a state of uncertainty (*FL*, 103–4). In his turn, when Berina implies some jealousy of other women when she tells Artander about getting in a fight with another woman who is reportedly in love with Artander, Artander seizes the opening, driving home his advantage by implying his attraction to a woman nearby (*FL*, 110–11). The letters interact, commenting on each other by direct and indirect response, by overwriting episodes and reencoding messages from one correspondent to the other. If Berina and Artander eventually take away some understanding of each other, so too does Davys's audience for the epistolary form makes clear the ways in which each character attempts to dominate and be dominated in their relationship.

Davys also uses symbols through the correspondence to signal emotional development. Artander repeatedly urges Berina to provide the "news" she has gathered from tea tables. Like Eliza Haywood, Davys recognizes what Artander's request really means, as the tea table had been famously idealized by Addison and Steele as the proper place for female education, which was comprised of good manners and the news of the day.[33] When Berina's political views become too much for Artander, he asks her to write letters "fill'd with the innocent Diversions of the Town," that is, "a little *London* news: I mean, such as the Tea-Table affords" (*FL*, 101, 106). When Berina begins to capitulate she provides "Tea-table Chat," reporting initially that "The Town goes on as it used to do, full of Party, Pamphlets, Libels, Lampoons, and scurrilous Ballads" and eventually ending a letter because she is "interrupted by two or three Ladies who are just come in, and my Correspondence must give place to the Tea-Table" (*FL*, 109, 114). That use of the tea table signals a capitulation by Berina similar to Artander's acknowledgement of Hanoverian legitimacy ("since we have made him our King, we ought to use him like his Predecessors, and give him the Honour due to the Kings of *England*" [*FL*, 110]).

Davys uses the exchange at the end of the narrative as a form of contract negotiation. Artander's pivotal letter concludes by encouraging them both to capitulate: he promises that her "least Inclination shall be a Command" and follows that immediately by asking her to "comply" (*FL*, 117). Berina rejects his proposal and reframes the relationship. "[T]he Promise you make of inverting the God of Nature's Rules, and being all Obedience, is no Inducement to me to become a Wife; I shou'd despise a Husband as much as a King who wou'd give up his own Prerogative, or unman himself to make his Wife the Head," she informs Artander (*FL*, 117). Protesting that "We Women are too weak to be trusted with Power, and don't know how to manage it without the Assistance of your Sex," she explains that "The notion I have always had of Happiness in Marriage, is, where Love causes Obedience on one side, and Compliance on the other, with a View to the Duty incumbent on both. If any thing can sweeten the bitter Cup, 'tis that" (*FL*, 117–18). Hence, Berina not only speaks of political rights in almost classically Lockean, Whiggish contract theory terms, she also speaks of marriage in equally classic Lockean sexual contract theory terms, articulating what Carol Pateman points out is the unequal dynamic in the sexual contract.[34] She rejects Artander's extremes, that is, his Toryism early in the correspondence and his radical egalitarian protestations at the end (although she is perhaps right to distrust them, as the odds of Artander truly adhering to such principles seem slight).

This dynamic seems to be what Davys is aiming at with the correspondence. The dates and timing of the correspondence at the end of the narrative are structural clues to where the relationship is going and to what Davys means by it. Artander's proposal of marriage on January 7 gets an immediate response on January 8, to which he responds on January 9. This is the only time in the novel that the letters are so closely spaced together. The situation is urgent, to say the least, and yet the resolution seems ambiguous. Artander's proposal on January 7 gets an immediate response on January 8 in which Berina offers to find him someone else to marry; although he replies immediately, on January 9, the tone is more formal than passionate: "I Hope, Madam, I may take your last Letter for a Compliance with my Wishes, and believe you are already mine; since no body, but *Berina*, can deserve the Character you give the Lady I am to expect from your Hand" (*FL*, 118). How convinced is he? The verbs suggest not very: "Hopes," "may," "believe." Nor does the tone. Nor does he rush to her side. He promises to come to London in "A few Days more" on January 9, and when he gets a reply (three days later), it takes him two weeks to write again, this time that he really is coming. This is not the behavior of a confident man; it sounds more like his language and strategies in earlier letters, in which he commanded her to write on unthreatening topics and veiled threats in stories about other people.

What about Berina? Is she going to accept his proposal? Does she know, herself? (Does she know herself? one might also ask). Berina seems to be holding him off in this final exchange, rejecting his vision of marriage, offering him a different woman, and then waiting three days to reply to a letter from a man hoping that he may believe that she is going to marry him.

Like Jane Barker with her contemporaneous framed narratives, Davys uses epistolary form to call attention to the problem of evidence and knowledge: whatever the characters may know of each other and themselves, the reader knows something else, something more, and in this case is also required to judge.[35] Lindy Riley observes that Davys denies simple resolution with this ending; I contend that this ambiguity is the point.[36] Throughout the correspondence and up to the very end, each time one writer provides a weak spot or makes an admission, the other finds a way to turn it against him or her. In the indeterminate responses, the struggle for power in the exchange of letters, and the ostentatious absence of the conventional happy ending, Davys reveals the underlying problems of heterosexual relations, particularly the difference between what one might call a Tory view of marriage and a Whig view of marriage. After all, Berina's outline of the ideal conjugal relationship is the Lockean sexual contract to a T. The politics of the state are the politics of love, and Davys forces her readers to realize that sameness by watching those politics play out in the letters. Epistolary structure provides an outsider view, exposing undercurrents that the correspondents are not or not fully aware of and allowing Davys to reveal in turn the problems with romance.[37] Davys's epistolary structure thus creates a detached perspective from which to evaluate the values driving characters, society, and the state. It is full knowing, in other words, that requires—or exposes—morality.

Familiar Letters is structurally anomalous within Davys's *oeuvre*, as it is the only epistolary novel she wrote. This extended treatment is meant to show Davys's interest in structure as a way of establishing a framework of knowledge and a framework of morality, and her view of knowledge and morality as different aspects of human existence. Davys's solitary use of epistolary structure also shows that she was willing to experiment and able to change methods when one did not serve her purpose as well as another. This willingness appears in the structural adjustments Davys made in her other novels as well, from *Alcippus and Lucippe* to *The Accomplish'd Rake*. Over the course of her career, Davys experimented with narration in search of a structure that would sustain a point of view detached from events within the narrative but providing the interpretive and evaluative framework, the moral context, for understanding those events.

The Cousins serves as a structural Grand Central Station for the rest of Davys's novels. It uses framing devices like *The Lady's Tale* and *The Fugitive/Merry Wanderer*. It also provides a strong central narrative consciousness like *Alcippus and Lucippe*, *The Fugitive/Merry Wanderer*, *The Reform'd Coquet*, and *The Accomplish'd Rake*. *The Cousins* and the other narratives structurally connected to it also share with *Familiar Letters* an interest in balancing separation and integration. *The Cousins* was published in 1725 in the *Works* but probably written between 1700 and 1704 when Davys was in London. It was republished in 1732 as *The False Friend*.[38] Because *The False Friend* is almost identical to *The Cousins* and it is identical structurally, and also because there is no doubt that *The Cousins* was published under Davys's eye, this discussion references the earlier iteration of this narrative.

In *The Cousins*, Davys manipulates the framed narrative to investigate narration's possibilities for presenting a moral framework and reliable knowing. It is the story of a pair of cousins, Elvira and Lorenzo, recounted by a narrator whose connection to them and the source of whose knowledge of events and inner thoughts are never revealed. Occasionally, secondary characters tell their own stories as narratives within the larger narrative of the cousins. At some points, the narrator presents emotional conflict between characters through direct dialogue, as when Elvira rejects her father's best friend's amorous advances despite his threats, or when she teases Lorenzo before admitting her love.[39] Overall, it is a narrated tale interlarded with several sub-narratives presented as framed narratives.

The Cousins is preoccupied with separation, both thematically (its characters are constantly being removed from their communities and families) and structurally (its frame is entirely detached from events and characters). The narrative plays out a variety of scenarios of isolation and exile. Gonsalvo retires to rural isolation with the infant Elvira when his wife dies, refusing to see anyone except his friend Alvaro. Elvira leaves her loving father to live with her aunt and her struggle to return to him when she gets engaged provides part of the novel's tension. Her cousin Lorenzo has been abroad for several years. Brokenhearted Octavio roams homelessly. Brokenhearted Clara does the same. Villainies are perpetrated by men who are exiled or excluded in some way. Alvaro, who attempts to rape Elvira while her father is away, was once ejected from his father's home and his inheritance by a younger brother in a replay of *King Lear*. Although he has been reintegrated into his family and society by Gonsalvo, Alvaro is not part of Gonsalvo's family and is much older than Elvira's father, so he is still a figure of separation. On his deathbed, he confesses that he fully expected to become even more of a pariah after attempting Elvira's virtue: "I...sate me down to consider, where I shou'd go to hide myself from all that knew me. I was very sure my Folly wou'd spread itself all over the Country" (*Cousins*, 2:256). Sebastian, also a rejected, much-older lover of Elvira's, transforms from pest to homicidal stalker when he recognizes that Elvira not only has chosen another man but also has closed the family circle and affirmed a powerful endogamy by selecting her first cousin, Lorenzo. The narrator announces that "from that moment [he] secretly vow'd their Ruins: He was no longer able to sit a Spectator of their Happiness, and therefore took his leave with a Look which promised nothing but Destruction to all around him; and wished for nothing more than the Eyes of a *Basilisk*, that he might have darted his venom, and looked them dead before he left them; and thus, brim full of damnable Malice, he went to put his diabolical Designs in execution" (2:218). More even than Alvaro, Sebastian's murderous mutterings connect his outsider status to his villainy: "*If they live, I cannot,*" Clara hears him explain, "*For I have a Hell in my Breast when I think they are to be happy*" (2:250). He poisons himself soon after, exiling himself from humanity and grace.

The narrative structure permits a more positive investigation of the relationship between moral integration and the social detachment that enables a healthy independent viewpoint. If for Davys's older men separation is the mark of depravity, for her narrator separation is the mark of virtue, both as a teller of narrative and of moral truths. Davys's narrator is capable of recounting events without evaluating them. When Elvira learns of Sebastian's suicide, the narrator describes and explains but does not judge her reaction: "Elvira, 'tho' she did not delight in Cruelty, yet was she not much concern'd for the Fate of *Sebastian*, who had so industriously sought her Destruction, but was very glad he had put a stop to all her Fears" (2:252). The narrator often provides the moral framework for understanding events, however. Having recounted the grief that Gonsalvo feels on the death of his beloved, virtuous wife as a result of delivering the baby they both had longed for, the narrator observes, "[W]hich only serv'd to shew him and the World, the instability of human Nature, which is never at a stay, nor content with what it has, but is still greedily wishing for those things, the grant of which, proves very often its greatest Disquiet" (2:206). On other occasions, she contextualizes the characters' behavior through irony or humor: "a Servant came in with a summons to dinner, where *Octavio* eat not much Meat, but fed plentifully upon Sighs" (2:243).

Davys also occasionally shifts seamlessly between the narrator's point of view and a character's point of view. This is a technique found almost exclusively in *The Cousins* as Davys attempts to move between framing narrative and framed narrative, and therefore between the values and reliability of one and the other. For instance, the narrator begins the sentence but Lorenzo ends it: "Thus, *said* Lorenzo, *Leonora* made an end of her mournful Tale; and then I got up, and with Thanks wou'd have taken my leave" (2:229). This technique is not quite the same as Jane Austen's transitions between an authorial, narratorial voice and free indirect discourse but it serves the same function, to move between a narrator's "objective" perspective and a value-laden "impersonal figural representation," as Daniel Gunn puts it.[40] In such rare but striking moments Davys uses what Laura Buchholtz calls "morphing," a shifting through different perspectives over the course of a single sentence from one point of view to another that Austen also uses for similar purposes.[41] This ability to shift between perspectives within the space of a sentence allows Davys to separate her narrator from events, present the pathos of each framed narrative, and preserve a moral framework in which the behavior of the characters—and their own assessments of moral codes—can be evaluated. Davys does not use morphing very much in her other novels; in fact, she ostentatiously eschews that technique in *The Reform'd Coquet* and *The Accomplish'd Rake*, as I discuss. Nevertheless, Davys's use of morphing in *The Cousins* suggests an early attempt to mediate the transition from detached point of view to integrated point of view. It also suggests an early attempt to mediate the potential disruption of authority in shifting from a narrator who ostensibly has nothing at stake in the narrative and therefore is reliable to a narrator who has a great deal at stake (what if Elvira concluded from his story that Lorenzo was a cad?) and is therefore not entirely reliable. Here, again, although this technique is not a third-person omniscient narrator per se, it does suggest efforts to develop literary techniques that will establish a reliable, detached, morally and factually authoritative perspective that presents its material as truth.

The Cousins is one of three novels that Davys revised and republished during her lifetime: the other two are *Alcippus and Lucippe*, which was structurally changed in the second version, and *The Fugitive*, which was structurally unchanged.[42] In the case of *Alcippus and Lucippe*, Davys's shift from first-person narration to a quasi-framed narrative suggests where her narrative interests lay. *Alcippus and Lucippe* uses a first-person narrator: Lucippe, looking back over an unidentified period of time at how she came to be married to Alcippus. *The Lady's Tale* is a framed narrative. *The Fugitive* and its second incarnation, *The Merry Wanderer*, also republished in the *Works*, begin with a first person narrator but eventually become so dependent on framed narratives that they both seem to change from one narrative form to another. *The Merry Wanderer*, unlike *The Lady's Tale*, is not structurally different from its original so these novels are discussed together. Critics have often taken the narrator of *The Fugitive* and of *The Merry Wanderer* as Davys's self-representation, but Bowden points out that the historical record reveals problems with this assumption.[43] Put in the context of her *oeuvre* the witty, didactic first-person narrator appears to be a recurring device. Whatever autobiographical facts might or might not be infused into *The Fugitive* (or *The Merry Wanderer*, not to mention other first-person narratives such as *The Reform'd Coquet* and *The Accomplish'd Rake*), this study addresses what Davys uses the first-person narrator to do.

In this group of novels—*Alcippus and Lucippe*, *The Lady's Tale*, *The Fugitive*, and *The Merry Wanderer*—Davys experiments with the narrators as knowers: what each one knows, how she knows it, and how that knowledge and her act of knowing are represented. The narrator and protagonist of *Alcippus and Lucippe* is Lucippe, the only child of an aristocratic family. Having reached her fifteenth year and the apogee of beauty, everyone around Lucippe begins to think of her marrying, although she is not inclined to it. After hearing about the handsome son (Alcippus) of an ugly but very wealthy Dutch friend of her father's, Lucippe experiences the usual mysterious, physical longings that in amatory fiction indicate love. These longings are followed by a mysterious encounter with a young man who is, of course, Alcippus; assorted trials and miscommunications; and a conclusion brought about by her own resolute actions, including following Alcippus to Holland and criticizing him in public for his bad behavior. The pair is reconciled and the novel ends announcing that they returned to England and married. *The Lady's Tale* recounts the same adventure framed by the story of Lucippe, renamed Abaliza, telling her own adventures to her long-lost friend Lucy, who comments on her tale from time to time. That change suggests that point of view is a central concern for Davys.

The focus in *Alcippus and Lucippe* is Lucippe's experience of these events. Lucippe herself announces that "I shall, with all the

Care I can…do my self Impartial Justice too" and "speak with Indifference."[44] Even when other characters have the opportunity share their views, they do so in direct speech or as reported by Lucippe. "My Mother, who had nicely observ'd my Looks, during the time I talk'd with my Father, soon found how I stood affected, and told him she fear'd he had driven his Jest too far, and desired him to say no more on a Subject which might be, for ought she knew, very pernicious to me," Lucippe explains (21). When her mother comes to see her later, "*Lucippe*, said she, methinks I need not ask how you do, your Looks are so much mended" (25). The narrative is missing a sense of retrospection, however; the focus on Lucippe's experience means much less interest in processing that experience intellectually or morally. There are very few interjections from a later Lucippe reflecting on someone's behavior, for example, and what retrospective moments there are do not involve the consequences of anyone's actions. It is a peculiar point of view, a Lucippe telling a story that must have happened in the past but with no significant function for the act of looking back. Without a retrospective evaluative perspective, the moral compass of the narrative resides in the actions of the characters in the narrative rather than in a perspective external to those actions. Lucippe's determined behavior in response to Alcippus' precipitate and selfish flight to Holland is to persuade the male authorities in her life—her father and uncle—to let her follow Alcippus. When she finds Alcippus, she berates him for his stupidity. The decision to follow him, the actively persuading the older men, and the hostile confrontation is remarkable behavior in a fifteen-year-old, especially one in 1704. That her behavior enables a happy ending suggests that this is the endorsed behavior of the narrative. Significantly, it is also unanticipated: Lucippe does not reveal in advance what she will do when she finds Alcippus; she just castigates him when she tracks him down, and she does not comment in retrospect on the behavior. In a way, then, Davys is telling the story without telling the story; that is, she tells the story and allows the point of view—Lucippe—to stand in for the moral center of the narrative without incorporating a mechanism for examining or justifying Lucippe's position and values.

This is a maneuver that recurs throughout the narrative, especially with Alcippus. Like Berina, Elvira, and Amoranda, Lucippe is rational and self-possessed. She is not unemotional—her mysterious passion for Alcippus puts the lie to that, as do her furious imprecations in Holland. Alcippus is impulsive and out of control, rushing about the countryside, jumping on a ship to Holland, and several times flying into a homicidal rage. Lucippe, however, is always in control, depending on good manners to negotiate emotionally charged social situations, ordering a voyeuristic stranger (who turns out to be Alcippus) to go away, and providing Alcippus with pious good advice upon the death of his father. When Alcippus stabs her footman, she immediately realizes that he needs to go into hiding to save himself from the law. Even with her parents, to whom she is deferential, she is able to balance proper filial obedience with self-defense, arguing with both parents when they want her to get married. So detached is the narrator that the morality resides in the actions and outcome rather than in the person recounting the narrative.

Davys refuses to contextualize the narrative with the prefatory material, as well. The Dedication to Mrs. Margaret Walker says a great deal about saying nothing:

> I hope I shall escape the Censure of a Flatterer, since I have (contrary to the Custom of Dedications) omitted those Encomiums which I know to be justly your due, and which are so Obvious to all that know you, that I should only be Laugh'd at, for undertaking a Task, which requires a better Pen than mine to Finish. I shall therefore only beg of you to accept of what is in my power.... [45]

Bowden calls it "a non-dedication."[46] The preface effects the same maneuver, explaining that because the author hates a preface, she will say nothing in it: "I should have saved my self the Trouble of Writing a Preface, had I not known the expectation of almost all Mankind, which is very much disappointed without one; and will no more allow a Book Compleat without a Preface, than a Lady fine without a Furbeleau Scarf; or a Beau without a Long Peruke."[47] The images emphasize the role of absence rather than presence, repeating "without" four times in the first sentence. The rest of the Preface parades its emptiness, as well:

> *And it is purely to satisfie those, into whose Hands this small Piece may fall, That I have said any thing more than barely the Subject matter; For which I can say but little, only that as it is chiefly design'd for the Ladies, so the most reserved of them may read it without a Blush, since it keeps to all the strictest Decorums of Modesty. But as it it [sic] would look very ridiculous for a Person to exclaim against late hours, which he himself sits up all Night; So would it be the very Abstract of Impertinence, in me to say I dislike a Preface, and at the same time Write a very long one: To prevent which and the Censure of the Criticks, I shall only with Sancho Pancho [sic] give you an Old*

Prover, viz: Little Said soon Amended, and so have done with the Preface. (Preface, np)

Here, too, Davys's prefatory materials refuse to preface the narrative; in so doing, Davys establishes that a lack of context for the narrative is in fact the context.[48] In rejecting adornment—rhetorical, political, social, material—Davys establishes the aesthetic and the moral framework for the novel, but she does so by stepping outside the norms to reflect critically on them. With a dedication that is not a dedication, a preface that is not a preface, and a romance that is not a romance, Davys balances inclusion (social, literary, moral) and detachment or exclusion.

Davys's decision to revise the structure of *Alcippus and Lucippe* to create *The Lady's Tale* has significant implications for the balance of detachment and inclusion when it comes to knowledge. The narrative frame in the latter version establishes a moral frame in which behavior is to be evaluated and understood. It clarifies the morality in which, by which, and through which the narrative functions and should be understood. Davys establishes the idea of illumination through comparison using the characters and the interactions of the frame. In *The Lady's Tale*, Lucippe has been converted into Abaliza and she reunites with an old friend, Lucy, in the frame narrative. Lucy persuades Abaliza to tell her love story, and when Abaliza obliges, both women interrupt her tale to comment on it and disagree. Primarily Davys uses Lucy to question Abaliza's behavior. Lucy is natural and unaffected, Abaliza is fashionable and fake. "Why, Child," Abaliza tells Lucy, "you'll be hiss'd off the Stage of Life, if you set up now for Sincerity," and Lucy replies, "Believe me, *Abaliza*, I grieve to see that honest Heart of yours, give so largely into the way of the World; I once thought you cou'd have scorn'd the name of Sycophant" (*LT*, 2:124). The names also draw a comparison.[49] Lucippe's name in *Alcippus and Lucippe*, like the title, signals that she is an inverse of Alcippus. In the narrative his irrationality, impetuosity, and violence contrasts with her reason, careful action, and desire to make or retain harmony. The title *The Lady's Tale* shifts the focus to the fact of the story rather than who is in it, and with the framed structure highlights the context in which the story is told. As for the names, "Abaliza" is as foreign as "Alcipus," signaling that they are both part of a fantastic, strange other realm, but they are otherwise unconnected.[50] Davys contrasts the un-English, exotic "Abaliza" with the solidly English and familiar name "Lucy." The focus of the narrative is on the interaction, sometimes the contrast, between the values of the story-within-a-story ("Abaliza") and the values of the frame ("Lucy").

Davys thus resituates the moral cues in the original narrative, moving them from being integrated into the story to being primarily external to it. The ending—Abaliza marries Alcipus—is not really the ending, because the story actually begins with Abaliza telling Lucy the story and reflecting on the fact that she is both at the end of her adventures and at the beginning of recounting them: "But you desir'd to have the Tale of my Amour, which I will give you at large. That, *said* Lucy, will be very obliging, because I have hitherto had only an imperfect account of it" (*LT*, 2:124). The narrative ends with Abaliza breaking off mid-sentence to join her husband and son (2:201). Lucy and Abaliza often frequently obtrude the frame by interrupting Abaliza's narration. This technique contextualizes Abaliza's thoughts and actions within the narrative. When Abaliza writes to Alcipus after her mother's death, Lucy interrupts: "Now, *said* Lucy, if I may speak my mind, I think it did make a Fraction in Modesty, to bid a Lover come and welcome, whom you had never seen above once or twice." Abaliza's response, "Pugh...the worst you can make of it, is, that I was a little too forward; but I liked the man and his circumstances, I did not know how soon his Father might force him away, and was resolv'd a little trifling Decorum should not part us for ever," does not express much respect for the basic modes of decorum but does evince considerable respect for Alcipus' superior fortune (*LT*, 2:155). Lucy is not wrong; she enunciates the moral framework within which Abaliza and the text are supposed to operate. By separating a moral voice from the events, the social framework within which those events operate becomes prominent.

Davys uses the prefatory materials associated with *The Lady's Tale* to support the exportation of knowledge and morality from the framed narrative into the framing narrative. The preface to *Works* claims *Alcippus and Lucippe* was less contextualized than it really was, emphasizing how the work was initially "naked" and "abandoned" until reclaimed and refurbished twenty years later: "*But meeting with it some time ago, I found it in a sad ragged condition, and had so much Pity for it, as to take it Home, and get it into better Clothes, that when it made a second sally, it might with more Assurance appear before its Betters.*"[51] With such claims, Davys underscores the difference in the position of the moral element in the two works and the improvement such a repositioning provides. *The Lady's Tale* and *The Cousins* are both intended to infuse novels with morality by "reforming" their representation of love and passions so people may learn how to "regulate" them (Preface, 1:vi). The revisions are designed to render more clear and effective the moral stance and the didactic effects of the novel.

The changes to *Alcippus and Lucippe* thus suggest that between 1704 and 1725 Davys remained concerned with the value of detachment and being an outsider, particularly as that position enabled a viewer to present and preserve a moral system for understanding events. What seems to have changed, however, is the position of that outsider. In *The Lady's Tale*, although Abaliza recounts her own life story, she is temporally and physically detached from the time and place of those events. Furthermore, the frame narrative removes the moral center from Abaliza altogether, placing it partially in Lucy and partially in the space between Lucy and Abaliza, in their interactions. At the end of *The Lady's Tale*, for example, Abaliza recounts the happy ending of her "amours" with considerable satisfaction but adds, "But now, my *Lucy*, as you are my Friend, and none a better judge in such Cases, I beg you will tell me, wherein I have transgress'd thro' the whole Oeconomy of this Amour" (*LT*, 2:200). Lucy demurs at first but Davys has Abaliza press her request by couching it in terms of interpretation: "consider, my Dear," she says, "of what service it may be to me: for since you are resolv'd my Tale shall be made publick, I desire my Faults may be so too, if for no reason, but to prevent the criticizing World from making their remarks" (*LT*, 2:200). Abaliza is heading off the criticism that might ensue from readers encountering the narrative without a moral frame, readers who will perceive the work as morally rudderless and critique it accordingly. Here is the same anxiety about interpretation and the role of narrative expressed by Haywood in *The Tea-Table* and discussed in the previous chapter, and in fact Davys has Lucy proceed to list very readily Abaliza's errors of conduct.

Framed narratives help shape *The Fugitive* and *The Merry Wanderer* as well, but these novels are structurally more like *The Cousins* than *The Lady's Tale*. Like *The Cousins/The False Friend*, *The Fugitive* and *The Merry Wanderer* emphasize events happening in the overarching framing narrative rather than events and perspectives in the framed narrative, where the emphasis lies in *The Lady's Tale*. This combination of narrator and framed narratives distances moral judgment from the events being judged. Also like *The Cousins*, Davys thematically and structurally emphasizes separation in these narratives. Thematically, Davys begins with the trope of the traveler and then compounds the figure's separation. The narrator is an outsider on a number of levels. She is traveling; she is an Irish woman in England; she is a widow.[52] The narrator moves from house to house, staying with friends, kin, or acquaintances, none of whom has she seen in some time and some of whom are actually strangers to her. In her first attempt to stay with family, her cousin is from home and his miserly, unpleasant wife hosts the narrator, who dines out literarily on the experience as she cannot dine literally during it.[53] When she visits a farmer to collect money she has left with him, she asks him questions about his marriage to a shrew that reveal how little she knows about his life (*MW*, 1:181–83). Overall, the narrator moves through the physical, social, and moral landscape of England, visiting people from whom she is very different and to whom she is consequently able to bring a new and improved perspective and morality.

Davys also creates a narrator whose moral judgments are true and whose knowledge is reliable, but who often observes more than she interacts with and influences other people. The Irish widow passes judgment on her own experiences at the moment that they occur, preserving an ironic, detached perspective even as she is contending with selfish hostesses and hypochondriac husbands. When she twits the miserly housewife for boasting about herself, the widow is reflecting satirically in the moment, from a detached and critical point of view, on the woman's behavior. "Madam, you have done me a singular Favour in making me acquainted with your Worth; but in those Parts of the World where I have spent my time, most People, whose Lives are attended with so many Advantages, leave the recital of them for other Folks," she states (*MW*, 1:166). Similarly, when she criticizes her friend for believing in ghosts, that criticism takes place at the moment of the events in the so-called haunted house (*MW*, 1:192–93). Davys adds a second layer of detachment with the retrospective perspective, that is, the way that the narrator recounts events that happened in the past rather than the response of the narrator to those events in the past. When the Irish widow recounts how the audience responded to the Londoner's tales of prostitutes ("As soon as we had shown our pretty white Teeth with laughing at his Disappointment, he took t'other Dram, and then began his second Adventure" [1:185]), she is taking a retrospective perspective on those events even as she describes her own satisfaction in the moment. Davys compounds the critical detachment and outsider stance of the narrator with a retrospective perspective in *The Fugitive* and *The Merry Widow* to establish a distance from characters—including narrator—and events that allows Davys to present them within a framework for understanding those people and actions.

At the same time, Davys frequently uses framed narrative in *The Fugitive/The Merry Wanderer* to further detach the narrator from her materials and better integrate those materials into a moral framework. After a young woman tells the story of how she was seduced and abandoned as a teenager to explain why she is refusing an eligible suitor, the Irish widow gives her helpful advice, providing a perspective on the situation and a code of ethics within which the young woman can act (*MW*, 1:195–97). The Irish widow's moral framework is validated by its success: the couple happily marries "to the great satisfaction of all that knew them" and the thanks of the bride, who "sent for me," reports the narrator, "and told me what good success attended my Advice, and how near she was to Matrimony" (1:199). On another occasion, the Irish widow tells the story of her friend telling her the story of a handsome young lady who winds up marrying the man whose bed she got into by mistake at an inn (*MW*, 1:200–211). In this instance, the Irish widow does not comment on the tale directly but both repeats the evaluation of the framed narrative's narrator without commenting on that evaluation and provides an approving description of the couple in their married state (*MW*, 1:211). Davys thus uses framing devices in *The Fugitive* and *The Merry Wanderer* the way she uses them in *The Lady's Tale*, to provide a moral context within which the actions of the narrator and the other characters can be understood.

Although *The Reform'd Coquet* and *The Accomplish'd Rake* seem at first blush considerably different from these other novels, they have structural qualities and novelistic concerns in common. Like *The Fugitive* and *The Merry Wanderer*, they both use a lively, personable narrator who provides the moral framework in which the story should be understood. Like *The Cousins* and *Alcippus and Lucippe*, that narrator stands outside the action, possessing complete knowledge of events and character but having no role within the actions or decisions of the people she describes. In *The Reform'd Coquet* and *The Accomplish'd Rake*, Davys gives up the framing device and deploys only sparingly the epistolarity of *Familiar Letters*. It would be tempting to call these novels the zenith of her skills and perhaps they are. Given the dates of *Works* and the recycled nature of so much of the prose fiction in it, it certainly looks as if Davys's entirely new compositions of the 1720s—*Familiar Letters*, *Reform'd Coquet*, and *Accomplish'd Rake*—are her most subtle and skilled. Given her structural choices in the revisions of the novels in *Works* such as the conversion of *Alcippus and Lucippe* to a framed narrative, however, it is also tempting to wonder if Davys was perhaps still seeking a structure that served her purposes.

Critics have attributed the significant difference in polish between *The Fugitive* and a late work like *The Accomplish'd Rake* to Davys's familiarity with the theater, but it seems at least as much if not more indebted to her ongoing engagement with narrative structure.[54] Certainly, the novel and the stage did have a productive relationship from the former's beginnings, and Davys's later works employ techniques derived from or influenced by the theater. Nevertheless, Davys's dramatic output was limited: *The Northern Heiress* was performed in London and enjoyed a very lucrative third night performance but Davys composed only one other play, *The Self-Rival*, and that was never produced.[55] In comparison, between 1700 and 1727, Davys wrote seven novels and rewrote at least two of them. It seems quite likely that this immersion in the novel affected her technique. Regardless of the narrative we construct for Davys's motives, however, the set of structures remains. The narrators of *The Reform'd Coquet* and *The Accomplish'd Rake* balance inside and outside, detachment and engagement, disinterested observation and didactic commentary, showing Davys's ongoing commitment to these ideas and to using narrative structure to engage with them.

Davys's narrators in *The Reform'd Coquet* and *The Accomplish'd Rake* could not exist without changes to epistemology catalyzed by and epitomized by Locke. That is not to say that they are Lockean narrators, or that Davys read Locke. Nevertheless, they are products of an environment in which ideas about detachability and the self are in play. Davys creates a narrator who has a body and a mind, but whose mind is not dependent on the body. Consistent with Lockean ideas about the self, the body is important to Davys's narrators. It establishes their authenticity and it creates a house for the mind. The body's experience in the past makes knowledge in the present—the narrated story—possible. Occasionally and more obtrusively in *The Accomplish'd Rake* than in *The Reform'd Coquet*, that materiality is invoked in order to explain how the narrator's knowledge is acquired. Even when extravagantly improbable—"of which more hereafter," she announces, "for I am this Minute going to take Coach for *London* again, where I left my young Knight" (*AR*, 154)—the materiality is used to explain the experience that enables knowledge. But that is all that the body does for Davys's narrators in *Reform'd Coquet* and *Accomplish'd Rake*. The rest is carried by the mind or consciousness, which provides the narrative facts and the moral framework of the tale. The narrator's understanding shapes the audience's understanding, because her understanding provides the insights that guide interpretation. Consistent with ideas from natural philosophy that were crystallized in Locke's *Essay*, the detachment of the mind from the body, and the detachment of Davys's narrator from what she has observed, define the reliable, knowledgeable self and thereby ensure reliability.

The Reform'd Coquet opens by summoning the presence of a narrator, as the first two sentences of the novel are a self-description: "The most avaritious Scribbler that ever took Pen in hand, had doubtless a view to his Reputation, separate from his Interest. I confess myself a Lover of Money, and yet have the greatest Inclination to please my Readers; but how to do so, is a very critical Point, and what more correct Pens than mine have miss'd of."[56] Shortly thereafter, the narrator illustrates a didactic point—"Vanity, which is most Womens Foible" often gives rise to other vices—with a brief story of her own vain sister. This short autobiographical exemplum provides the narrator with a father, older sister, and disappointing brother-in-law (RC, 14). *The Accomplish'd Rake* follows the same strategy: "Young *Galliard* who is to be the Subject of the following Leaves, will (with his own Inclinations, and a little of my additional Discipline) be a very exact Copy of the Title Page; for tho I shall be very punctual in delivering nothing but plain Fact in the fundamental part of his Story, it is not impossible but by way of Episode I may intermit now and then a pretty little Lye, and since it is to be both little and pretty, I hope my Reader will excuse me if he finds me out, and let him convict me if he can" (*AR*, 127). Both openings provide a sense of a "there" there, a sense of a narrator's material as well as intellectual and emotional being. The narrator of *The Accomplish'd Rake* also has a body, complaining ironically that "Sure unjustly are we called the weaker Vessels, when we have Strength to subdue that which conquers the Lords of Creation, for their Reason tyes them down to Rules, which we like *Sampson* break the trifling Twine and laugh at every Obstacle that would oppose our Pleasure" (*AR*, 129). She also invokes her own presence in the telling of the story, although not in the events in the story, at points when the shifting point of view seems to need some explanation. "[R]ather than lose so much Time, I think fit to return into the Country, and see how Things are transacted at *Galliard-Hall*, where I no sooner entered, than I saw *Tom* and his Wife arrive," she explains (*AR*, 148). Later, the narrator changes location with "of which more hereafter; for I am this Minute going to take Coach for *London* again, where I left my young Knight" (*AR*, 154).

Whatever experience her material existence has brought such a narrator, however, it does not appear to have much influence on her perception of events or characters. Her body makes her real but does not shape her knowledge beyond conferring the ability to acquire it. In comparison with the narrator of one of Aphra Behn's narratives, for example, the narrator of *The Reform'd Coquet* or *The Accomplish'd Rake* is not influenced by her body. The narrator of *Love-Letters Between a Nobleman and His Sister* could not recognize her own attraction to Octavio, for example, but no such problem attends the narrator of these novels. Davys uses the sense of a body or a material existence (why else take a coach?) to provide authority and authenticity, and to show the narrator's position vis-à-vis her characters: observing them, able to see and know anything she wants, but not interacting with them.

Once the narrator is grounded in materiality Davys can credibly, or at least not *in*credibly imbue her with a remarkable range of vision and knowledge. Reporting Formator's anxieties about Berintha and Arentia, she knows exactly what he is thinking and feeling without his saying anything aloud:

> *Formator* came into the dining-room, and with discompos'd looks, walk'd a few turns about it, saying to himself: From whence proceeds this strange uneasiness? Why is my Heart and Spirits in such an agitation? I never was superstitious, and yet I cannot forbear thinking *Amoranda* in some new Danger; there must be something in it, and Heaven in pity to her, gives me warning: Then after a little pause—I'll take it, *said he*, and watch the lovely Charmer: I know not why, but methinks I tremble at the thoughts of those two Women, and fancy I see her more expos'd to ruin now, than when she was surrounded with Fools and Fops. (*RC*, 53)

After Callid and Froth are defeated by Amoranda and Formator (and some burly footmen), the narrator knows that "tho' they liv'd like Scoundrels, they went off like Gentlemen, and the first Pass they made [in a duel], they took each other's Life" (*RC*, 33). In *The Accomplish'd Rake* the narrator reports that "*Belinda*'s heart was now restored to its former Quiet, and her Fear and Anger were both banish'd, for she saw the Looks of the Knight so much alter'd that she no longer doubted but his designs were so, too" (*AR*, 194). When Sir John learns that Nancy is pregnant, the narrator supplies his internal monologue: "O *Galliard*, said he, wretched *Galliard*, what has thou done? And how hast thou for a few Hours of brutal Pleasure entail'd an Infamy upon a whole Family, nay upon a Family that always loved thee in spight of my own Demerits" (*AR*, 176). The narrator explains that Lady Galliard, in turn, spends an entire night internally plotting to indulge her pleasures and later is troubled by "her own inward Accusations" (*AR*, 141, 152). Other characters, sometimes the characters themselves, do not know what the narrator knows about their inner workings, but because the narrator's corporeality renders her authentic, that intimate knowledge is much more acceptable.

The narrator also overtly provides the moral code within which the characters and their actions should be judged. Davys was forthright about her agenda throughout her career. "*My Pen is at the service of the Publick*," she announced in the Preface to the second volume of *Works*, "*and if it can but make some impression upon the young unthinking Minds of some of my own Sex, I shall bless my Labour, and reap an unspeakable Satisfaction*" (Preface, 2:6). But as De Bruyn also observes, the moral credibility of the narrator depends on her (sometimes ironic) separation from events, and the didacticism integrates with the narrative precisely because the narrator who provides didactic commentary does not.[57] The narrator is not part of the cast list, so her pronouncements and the value system within which those pronouncements operate thus appear to be authoritative and reliable. Detachment confers moral authority: it is precisely because the observant judge has nothing at stake in presenting or evaluating events that her judgments may be accepted as truth. Her stance—outside the interactions that she observes—confers accuracy in reporting and evaluating. In this way, Davys not only captures the philosophical ideal but also creates in rhetoric and structure the representation of this ideal and the depiction of it in action.

At the broadest level, the narrator provides general moral statements and therefore outlines the moral principles of the text. In *The Reform'd Coquet* she exclaims, "What an unhappy Creature is a beautiful young Girl left to her own Management, who is so fond of Adoration, that Reason and Prudence are thrust out to make way for it" (21). A "beautiful young Girl left to her own Management" might not seem to be such an unhappy creature and certainly Amoranda doesn't seem all that miserable at this early stage in the narrative, but the narrator's statement insists on reevaluating Amoranda's condition. She can be an "unhappy Creature" without knowing it when deprived of her "Reason and Prudence." The narrator of *The Accomplish'd Rake* makes similar assessments. "O Men of Merit say, what avails good Sence when left in the Hands of a careless Libertine, who had much rather tye it down with Links of Iron than listen to the Friendly Admonitions it kindly offers," she laments (156). More to the point of the novel's concern with education rather than birth, the narrator criticizes the problems with Sir John's upbringing in general terms:

O Man! how strong are thy Passions, how exorbitant thy Desires, and how impotent thy Virtues? Here have we a Person of Birth, of Fortune, of Sense before us, a Man, who might have been a Credit both to his Country and Species, had the early Rudiments of that Behaviour, which makes us value one another, been timely instilled while his tender Years were capable of Impression; but alas! the Want of Care in his Education made him a Perfect *Modern Fine Gentleman*; which, when we consider the sad Ingredients, they make a very Woful Compound: It is true, if we abstract bad Actions from Folly (which in my humble Opinion can hardly be done) Sir *John* was very free from the Imputation of a Fool, but then he had a double share of the Rake to make up his *Quantum*, and finish a very bad Character. (*AR*, 166)

It is worth noting that Davys here uses a vocabulary drawn from chemistry to make her point. "Compound" was expanding from a primarily mathematical usage to include chemical usage, as the eruption of medical books on making medicines at the beginning of the eighteenth century testifies. By the mid-1720s, the word's application in chemistry was well-established. The same was true for "abstract," to extract something from a compound, and "quantum," meaning something that was measurable.[58] "Ingredients" could also be a term from chemistry although it was used in other discourses, as well. Davys's imagery from natural philosophy, specifically from experimental chemistry, again testifies to her familiarity with the discourse and her view that it was useful for communicating with her audience. The idea that humans are a compound certainly sorts well with the distinction that Davys makes between the value of the corporeal and the value of the intellectual and moral in her construction of a narrator.

Davys's narrators also directly interpret characters' behavior and motives. "My Lord, with a little too much freedom, snatch'd her to his arms, took a Kiss, and vanish'd," the narrator states in *The Reform'd Coquet* (*RC*, 29) and points out that because Amoranda "had a Soul capable of Improvement, and a flexible good Temper to be dealt with, [Formator] made no doubt but one day he should see her the most accomplish'd of her Sex" (*RC*, 33–34). The narrator characterizes Lady Galliard's activities as "Faulty Pleasures" pursued "in the Criminal Company she best liked" (*AR*, 148). Even when the tone is sardonic, as it often is in *The Accomplish'd Rake*, the narrator's interjections, observations, and adjectives serve to explain or evaluate what she sees and reports.

This commentary often works in tandem with the moralizing provided by other characters. Altemira warns Amoranda that "a Man, who is resolved to be a Libertine, has no true value for a Woman's good Qualities" and the reformed Lord Lofty admits that "there is certainly a secret pleasure in doing Justice, tho' we often evade in it, and a secret horror in doing ill, tho' we often comply with the temptation" (*RC*, 43, 49). As such declarations indicate, this synergy between narrator and character is unsubtle: the primary moralizing forces of *The Reform'd Coquet* and *The Accomplish'd Rake* are named Formator, Teachwell, and Mr. Friendly, after all. When Sir John learns that Lady Galliard is having an affair with her footman, he "cr[ied] out with transport, tell me Mr. *Teachwell*, for you know the World, tell me I say, are all Women such?" and Teachwell not only answers him, but offers the reading and thinking skills necessary for living in society. "No, Vertue forbid, one single Faulter should infect the whole Species," he answers. "Women no doubt, are made of the very same Stuff that we are, and have the very same Passions and Inclinations, which let loose with a Curb, grow wild and untameable, defy all Laws and Rules, and can be subdued by nothing but what they are seldom Mistresses of" (*AR*, 137). Distinguish the specific from the general and recognize the role of context on character formation: one primary purpose of Davys's own narrative, one might argue.

This interaction between the framework promoted by the narrator and the characters' actions and statements is used particularly effectively to establish an opposition between reason and passion, intellect and appetite that appears throughout both *The Reform'd Coquet* and *The Accomplish'd Rake*. Good-hearted but confused Lady Betty in *The Reform'd Coquet* no sooner sees her brother remove his disguise then "she thought herself in some inchanted Castle, and all about her Fiends and Goblins" (*RC*, 83), an explanation that reaches for the realm of fancy rather than the realm of reason. Formator complains to Amoranda, "Madam, of what use is our Reason, if we chain it up when we most want it? Had your's had its liberty, it wou'd have shown you the villainous Designs of your Noble Lover" (*RC*, 30). Although Formator's reaction to Berintha and Arentia is physical and emotional, it is also rational. "Observation puts a great many things into our heads," he says to Amoranda, before enumerating the reasons why he thinks Berintha is a man. And while Amoranda refutes him with what she calls "several Reasons," they are based on assumptions about people's motives rather than observations of people's behavior, a judgment system that repeatedly fails her through the novel (*RC*, 54).

Amoranda's reformation consists in learning to resist the allure of emotion and pleasure and to seek instead the virtues of moderation, modesty, decorum, and reason. The same transformation

is incompletely achieved in *The Accomplish'd Rake*, as Sir John
(and Dolly, his sister, and even Lady Galliard to some extent) initially learns to pursue individualistic hedonism and at the very end
attempts socialized, rational virtue. The narrator contextualizes Sir
John Galliard's behavior as sensual and in opposition to the reasonable behavior of virtuous people. After raping Nancy, "Sir *John* was
now again at *London* entertaining his Five Senses with every Modish
Delight" and he tells himself to bury his remorse in libertinism (*AR*,
170). "This sensual Soliloquy," the narrator observes, "set our Knight
upon searching after new Pleasures" and opposes the "secret Impulses
of his Mind (which he was very loth to call Conscience)" (*AR*, 171,
170). Like Amoranda, Sir John resists reason in favor of the irrational—hedonism, vanity, the exercise of unbridled power—and must
be brought to love it instead. Unlike Amoranda, however, he is not
"Reform'd" but "Accomplish'd" and his hold on reason and virtue is
tenuous, as the narrator notes (*AR*, 226).

The narrator is particularly present at the ends of the novels. She
shows Nancy's decision in *The Accomplish'd Rake* to marry Sir John
as a very practical and illusionless one. Having discovered that the
man proposing to her is also her rapist, "The poor Lady trembled
with Resentment, but recalling her Temper said as follows: Your barbarous Usage Sir *John*, might very well countenance a firm Resolution
of seeing your Face no more, which I should certainly make were
I only to suffer for it, but I have a Child which is very dear to me,
and in pity to him I will close with your Proposals," after which she
demands a social and financial arrangement to the child's benefit
(*AR*, 225). That practicality is reinforced by the narrator, who notes
that Sir John's willing compliance and reform was hardly to be relied
upon for any length of time: he "was now resolved to hasten his new
Design least a returning Qualm should rise to stop his generous and
honourable Intentions" and the narrator "here tell[s] him that I have
set two Spies to watch his Motions and Behaviour" so she can expose
him "by Way of Advertisement to the Publick" if he backslides (*AR*,
226). After all, he is no reform'd rake as Amoranda is a reform'd
coquet. He is an accomplish'd one.

Davys's concluding arrangements for her characters are not only
cynical and skeptical but also open-ended. *The Reform'd Coquet* ends
with a speedy marriage and an equally prompt departure for London,
hallmarks of heterosexual normativity and conservative gender politics. Davys uses the narrator to destabilize this ending, however, by
opening it up: the characters "went to *London*; where the Reader, if
he has any Business with them, may find them" (*RC*, 84). The ending

is not really an ending; the wedding is an endpoint only because it is expected to be the endpoint. Davys does the same thing at the end of *The Accomplish'd Rake*, although more extensively. In addition to the marriage of the protagonists, novelistic convention urges the marriage of Nancy's brother and Sir John's sister, but the narrator resolves to "leave their Story to that grand Tell-tale old Father TIME to begin and finish" (*AR*, 226). Nor is Sir John's reform guaranteed. The narrator speaks to Sir John, warning him that "I have set two Spies to watch his Motions and Behaviour, and if I hear of any false Steps or Relapses, I shall certainly set them in a very clear Light, and send them by Way of Advertisement to the Publick" (*AR*, 226). No one wins and everyone knows it.

This open-endedness allows Davys to call attention to the artificiality of narrative and therefore of its values. As De Bruyn observes, *The Reform'd Coquet* and *The Accomplish'd Rale* repeatedly expose the formulaic conventions as just that: conventions and expectations.[59] The narrator underscores how a marriage between Nancy's brother and Sir John's sister is consistent with narrative expectation and patriarchal expectation but not with human nature by her refusal to include it in her own narrative (*AR*, 226). "As for Sir *John Galliard*," the narrator states at the end, using his full name to suggest the created character rather than the sense of a person, "I would have him acknowledge the Favour I have done him, in making him a Man of Honour at last" (*AR*, 226). By doubting that Sir John can maintain his reformation, the narrator evokes the disjuncture between narrative convention and human behavior: whether he can sustain even his dubious reformation, whether surveillance can keep him in line, whether the narrative—*The Accomplish'd Rake* or a reform narrative—can achieve closure. The same applies to the more conventional ending for *The Reform'd Coquet* (*RC*, 84). The narrator thus functions to call attention to both novels as "made things."

What is Davys's point in highlighting these novels as fictions? Susan Glover argues that Davys' "subtle and open-ended questioning" and her "complex, subversive, and acute rendering of gender and property issues" defy the notion of the didactic.[60] The period construed the didactic more broadly than this view would suggest, however. In the case of the open-ended endings, Davys calls attention to the act of narrativizing as the creation of a value system. She deploys two strategies to expose her narratives as collections of conventions. She uses the detached narrator to comment on the narratives as artifices and she absents her narrator at key moments to force readers to judge conventions as conventions—of narrative, of gender, of relations between men and women. Davys signals the craftedness of the narrative with phrases such as "I will now leave them a while to compare Notes together, and step back to the *Bagnio* to see what becomes of the two Antagonists" (*AR*, 162), phrases that call attention to jumps or changes in narrative and perspective. The embodied narrator may be authoritative, but she is giving the audience the fruits of her labor. This attention to the narrative's artifice goes beyond the episodic. If *The Accomplish'd Rake* ultimately contains Sir John and Nancy Friendly within the conventions of romance, for example, the narrator's commentary at the end shows that enclosure to be highly problematic. Their marriage is forged in the crucible of social conventions and while the reabsorption of Nancy into society is helpful to her because it benefits her child, the event is not also a happy one. Nor is it a happy ending for Sir John, who marries a woman he pities and becomes subject to perpetual surveillance and the attendant threat of public shame. Similarly, the novel flat-out refuses to marry off the other potential couple, recognizing how novelistic convention is entirely inadequate for the realities of character, however convenient it might be for narrative and for the circulation of property. Marriage may be the only solution for exile, but it is no boon. If the ending is conservative, then, it is conservative because it recognizes where power lies, not because it endorses that situation. Glover makes a similar point about Davys's handling of property: "the secure control of the power of landed property over the lives and wombs of the women in her fiction is challenged in the very act of its acknowledgment."[61] Davys's refusal to endorse social norms even as she acknowledges their power raises questions about the nature of her conservatism and suggests that such a classification is insufficiently nuanced to describe her views.

The differentiation between the narrative and the narrator, effected through the narrator's separation from the narrative's events, also allows Davys to "absent" her narrator from parts of the narrative. Davys uses an absent narrator the same way she uses epistolary form in *Familiar Letters*: the evaluation of the characters and events appears in the juxtaposition of speech and action, provided by an entirely separate point of view whose moral view has created the context in which the characters and events appear. When Amoranda attempts to question Alanthus's refusal to rescue her from Biranthus in *The Reform'd Coquet*, for example, the narrator remains outside their conversation: "I own, my Lord, *said* Amoranda, I owe a thousand Obligations to your generous care, and my whole Life will be too little to thank you for them, but—No more, Madam, *said he*, *interrupting her*, I had a glorious return for all that Care, when at Night, as *Formator*, I heard the whole story over again, and so much in favour of the happy Stranger, as *Jove* himself would have listen'd to with envy" (*RC*, 81). He cuts her short at "but," at the moment when she would have pointed out a flaw in his behavior, thus silencing her criticism. The interruption construes her statement as thanks and her emotional state as gratitude, although in fact her opening point is sincere but also a requisite courtesy and a softening technique. When Amoranda then immediately points out that he used the knowledge of her partiality to torture her further by sending word that he was ill, an action she calls "another piece of cruelty to lay to your charge" (*RC*, 81), his response to this accusation is equally selfish. "My dearest *Amoranda, said he*, pardon that one tryal of your Love, it was not possible for me to deny my self the exquisite pleasure I knew your kind Concern wou'd give me," he answers, before turning the subject to his own suffering, exclaiming, "but good Heavens! how did my longing arms strive to snatch you to my bosom when you had read that Letter, that I might have suck'd in the pleasing tears which drop'd from your lovely Eyes" (*AR*, 81). Throughout this exchange, the characters speak directly. The only material supplied by the narrator is description: "said he, interrupting her" and "This call'd a blush into *Amoranda*'s Cheeks" (*AR*, 81). The interpretation of the exchange, the "spin" on Alanthus' behavior and motives, is provided by Maria, who puts her seal of approval on Alanthus' behavior by telling Amoranda that she wishes someone would speak to her so nicely. But Maria is a character and therefore part of the object lesson; furthermore, she is a spinster (she admits to being over thirty as well as unmarried) and is also being reformed to desire love and marriage, which hitherto she rejected (*AR*, 72). The detached narrator shows the characters conspiring against Amoranda, thus revealing the social pressure and its curbing, silencing effect for the audience to consider. Amoranda is not the only didactic object; the context in which she operates is, as well.

The narrative thus exposes the actions of the characters as well as the evaluative framework used to interpret those actions. By requiring the conventions of the narrative to stand on their own, Davys highlights the value system within which the novel is concluded. Amoranda cedes her independence by siding with Alanthus and Maria against Lady Betty when they continue the charade of Alanthus's disguise, shocking the young woman so badly that she faints (*RC*, 83). At this point, Amoranda's full integration into patriarchal romance is complete; although she avers that "the Scheme of the Beard...I had no hand in it," she has gone along with the game at the end and she never has direct speech again (*AR*, 83–84). In effect, she vanishes. It is worth noting that Mr. Traffick is the only character who immediately recognizes Alanthus under the Formator disguise; as each woman gains the knowledge that the old teacher is the young man, she is captured by the patriarchal order and silenced, her independent motion (like Lady Betty's traveling about the county, like Maria's journeying where she pleases) limited (they all go to London together and stop moving).[62] Entering into knowledge of the male's importance is symbolic entrance into patriarchy, a point presented without commentary by the narrator and enabled by the narrator's separation from events. Spencer reads this ending as Davys's capitulation to male authority, but it is possible to read it as a critique rather than an acceptance of the distribution of power—the power of society, gender, and narrative convention.[63]

Davys absents the narrator so the juxtaposition of events and behavior exposes the text's values in *The Accomplish'd Rake* as well. Susan Staves points out that in *The Accomplish'd Rake*, this "tactic" allows Davys to "characterize[...] undesirable male characters by allowing them to pontificate foolishly on their mistaken ideas about women's nature."[64] This method also serves broader purposes. When Belinda rejects Sir Combish's offer of marriage, the scene unspools in direct dialogue (*AR*, 208–9). "You know, Sir *Combish*, our Passions are not at our Command; and if we hate when we should love, it is owing to a Depravity in our Fancies, which we may strive against, but can seldom master: This is just my Case," she tells him, to which he replies, "Did you really say, you could not accept of my Offers which is Honourable Love; and what at first I did not design, and perhaps more than some People deserve[?]" (*AR*, 208). His retort reveals his intentions, his character, and Davys's larger point about the barbarizing effects of being an English gentleman. Like in *Familiar Letters*, for example, part of the function of withdrawing the narrator is to expose the inequities of the social system. Davys uses the balance of detachment and engagement made conceivable by Lockean notions of the self to investigate the nature of narrative, including the role of observation in narrative, and narrative's relationship with morality.

The detached narrator also allows Davys to question gender norms and romance through character as well as through plot and episode. Davys's women struggle to meet with their equal even if they are unaware of that struggle. From the beginning of her writing career, Davys's narrative prose rejoices in the nimble, cutting wit characteristic of Restoration comedies and Augustan satire.[65] Unlike Millamant and her sisters, however, Davys's women rarely have a sparring partner worthy of their powers. Artander repeatedly attempts to control Berina's speech in *Familiar Letters*. Amoranda bests most of her suitors with words—"Hang the fish, *said my Lord*. Aye, *said* Amoranda, for we shall never drown them" (*RC*, 17)—and Belinda not only halts Sir John's efforts to rape her with a well-aimed verbal blow but also spars with every man who approaches her (*AR*, 193–94). Like Lucippe's indictment of Alcippus in *Alcippus and Lucippe*, *The Accomplish'd Rake*'s low-born Betty corrects her former fiancé William with a furious, accurate assessment of his perfidy (*A&L*, 102–4; *AR*, 206). In addition to the repeated contrast of the rational, thinking woman with the impulsive, irrational, or sensual man, the characters' unmediated actions contrast with each other and expose the values and dynamics of the society in which they live.

The "disappearing narrator" thus separates the evaluative component from the observational component and the fact of the body from the functioning of the mind. These distinctions are fundamental to the epistemology developing through the philosophical revolution in England during this period, as I discuss in chapter 4. These notions depend on the idea that the self can be detached not only from its own body—from its sensory functions, in particular—but also from its context. As chapter 1 discusses in more detail, the interest in establishing a reliable relationship between body and mind, and self and context, can be traced at least to Francis Bacon and *The New Organon* (1620) and was significantly articulated by Locke, whose work became an epistemological flashpoint from the late seventeenth century on. Davys's novels show that it was not necessary to have read Locke to be interested in such ideas about the self and knowledge, and that ideas about detachment, knowledge and morality continued to influence narrative innovation decades later. Anxieties about the detachment of moral value and the body such as those expressed by Aphra Behn's *Oroonoko* had lessened considerably through discussions of Locke's separation of interpretation and observation from the body. And while it is difficult to imagine how the disembodied, authoritative narrators of Austen or Scott or Dickens might have evolved without Locke's contributions, as Lorraine Daston and Peter Galison point out by tracking the development of objectivity over the century prior to Austen, such narrators are not simply the result of Locke's work, but the outcome of decades of writers grappling with ideas of what it means to know something and how that knowledge is acquired.[66]

The new epistemology may have liberated knowledge and authority in some ways, but Davys's inconclusive endings suggest that it had neither finished the task nor made a neat job of it by the end of the 1720s. Davys's disinterest in the role of the body's sensory processes seems to sort ill with her concern about women's position, authority, and knowledge in society. If being outside society offers a valuable perspective, then what might that mean for women, who are outsiders in many respects, or for the body of the woman? On the other hand Davys's good women, even some of the less-good ones, are certainly powerful characters. Furthermore, Davys also seems acutely aware of gender's role in providing or eliminating opportunities, such as when Artander and Berina conduct their epistolary duel, when protagonists like Alcippe struggle with an inconveniently impulsive partner, or when a failure to be fully rational and critical endangers women like Amoranda. Davys's conjugal conclusions seem to suggest that for Davys, marriage may be the inevitable and the best narrative available but "best" is a relative term, not an absolute. Ultimately, Davys's narrators are never wholly disembodied, her narrative structures never entirely detached from the sense that there is a "there" there. Whatever benefits that Davys's narrators suggest might accrue from being detached from systems of power or from the primary narrative of society, her female characters suggest that there are also perils to being separate. Perhaps most significant in Davys's narrative structures, then, is that until the end of her active writing life Davys was still investigating and exploring, her technique and the ideas underpinning it still evolving. Like epistemology and philosophy at the end of the 1720s, Davys's narratives and the novel were still in process.

Davys's novels, particularly when seen in the context of other novels being written during the early years of the genre's emergence, show how early novelists were grappling with ideas about the self and knowledge with origins in the philosophical revolution and its ongoing debates and discussions. The work of Jane Barker, Eliza Haywood, and Mary Davys reveals that by 1730, there was no tidy ascendance of any kind of thought, not even Lockean or Whiggish thought, in the novel or by extension, in the society within which and in dialogue with which the novel was developing. Rather, the novel's narrative and rhetorical experiments provoked by ideas in circulation—ideas about what humans are capable of, about what knowledge is and how it is found or made—reveals ways in which the novel and concepts such as the self and knowledge came into being. Ideas about stability, unity, reliability, and completeness that underpin the broad category "literary omniscience" were all in play during this early period of the novel's development, a phenomenon that both provides the terms that we gather under "literary omniscience" and explains why that category is so inclusive and shapeless. Omniscience is the performance of something. The novel is as well: the performance of what we think it is to be human and to do the things that humans do, like knowing. The period between 1660 and 1730 saw the creation of many of the conventions of that performance, and with them, the conceptual and literal language that we use to tell the story of us.

Conclusion

Jonah Lehrer observes that "Every brilliant experiment, like every great work of art, starts with an act of imagination."[1] Aphra Behn, Jane Barker, Eliza Haywood, and Mary Davys would have agreed. However credible natural philosophy's productions, the English philosophical revolution's methods and epistemology depended on an imaginative act: a reconceptualizing of the human self. As *Women, the Novel, and Natural Philosophy* argues, it is the same imaginative act underpinning the development of narrative structures forming a new genre, the novel. Changes to ideas about the self grounding the seventeenth- and eighteenth-century revolution in natural philosophy were taken up by novelists interested in the same questions about what humans could know and how that knowledge could be possible.

Philosophers of the revolution held, implicitly and explicitly, that the stable, unified self, albeit intellectually and perceptually unreliable, could also compensate for that unreliability through intellectual training and particular methods, emotional and physical detachment, and the use of tools such as telescopes and air pumps. That compensation could yield a previously unimaginable breadth and depth of knowledge about the workings of the natural world, one that to some thinkers of the time posed a challenge to religion and to other thinkers provided its confirmation. But under it all, requisite to the new epistemology, was the new self.

This new self proved difficult to accept in some quarters, as revealed by debates about John Locke's *Essay Concerning Human Understanding*, perhaps the age's most powerful description of it. Even as late as 1727, the year of Isaac Newton's death, that self was the subject of contention and investigation. Furthermore, the idea of the self had crossed the permeable membrane between natural philosophy and popular culture as early as the 1660s, generating debates and discussion within popular culture or moving into popular culture predating Locke and following his seminal publication. The questions such a notion of selfhood provoked and the new rhetoric required for writing about it put tremendous pressure on traditional forms of written expression. The novel, emerging as a genre from riffs on extant genres and textual responses to other socioeconomic and political pressures, proved a useful site for engaging with questions about the self and with the rhetorical opportunities those questions afforded. Novelists, male and female, took up the opportunity and the challenge.

Women authors like those discussed here recognized that women had a lot at stake in this new notion of the self. It did not include them, first of all, and so systems and institutions predicated upon that self—the family, the state—provided little or no room or role for women. Furthermore, associations of morality and women rendered questions about whether a detached, autonomous self could be socially and morally viable of particular interest to women. That is not to say that women alone wrote about the self or innovated methods for representing or challenging notions of the self. Nor is it to claim that all women writers felt the same about these ideas about the self—as the different chapters show, their views were unsurprisingly diverse. Aphra Behn was highly skeptical that a person could be clear-sighted about himself or others, and thoroughly convinced that language was too deliciously unstable to permit anything but the manipulation of performance. Mary Davys, writing four decades later, was also sensitive to the artificialities of literary convention, but her narrative structures do not question the possibility of narrative or personal detachment, only whether and how detachment can be infused with moral value. Under the pens of such authors, gender—doubtless in addition to others I have not delineated—colors the treatment of the self in ways that make its problems and possibilities more visible and acute.

Just as their attitudes toward the self were complex and diverse, so too were the strategies such authors developed to address their concerns and enthusiasms. Behn exploited the possibilities in language and narrative structure to challenge the idea of the unified self capable of compensating for its flaws and achieving reliable knowledge. For Behn, language's own instability and malleability became an Achilles Heel not only for writing about the self but also for the idea of the self. How could we ever understand ourselves or each other if the tool by which we were educated to think was so protean and unstable? Jane Barker's compound narrative structure suggests a greater comfort with the idea of a unified self than Behn's narrators, as well as an interest in the epistemological possibilities of aggregation. For Barker, more confident in developments in experimental philosophy than Behn, the self's fallibility proved a positive opportunity for narrative and epistemological innovation to compensate. Eliza Haywood, who was also adroit at the layered narrative technique characterizing Barker's work, and Mary Davys were both concerned with the ramifications of a detached and detachable self for society, morality, and women, but they too viewed positionality as less of a danger than Behn. For Haywood in *The Tea-Table* and for Davys throughout her novels, detachment offered significant opportunities as well as perils for creating moral selves and morally valuable knowledge. Autonomy and detachment remained double-edged swords, advantages and disadvantages rolled into one. Haywood's and Davys's narrative structures reflect this mixed nature and ambivalent attitude. All four authors recognized the contribution of language, especially text, to creating the sense or the illusion of self; all four saw that contribution as potentially both perilous and empowering; all four wrote narratives out of those interests.

This study depends on the recognition that narrative structure such as point of view has a strong historical component, whether the structure reveals pressures of the past that helped bring it into being or whether it is responding to pressures in its own historical moment. For the most part, narrative theory does not account for the role of history in the creation of literary structure, as critics such as Paul Dawson and Susan Sniader Lanser have pointed out.[2] In the case of theorizing point of view, narrative theory's insensitivity to "historical contingency" as Dawson puts it, means that it attempts to use a twentieth- or twenty-first century notion of self to theorize representations of a self constructed within a very different set of ideas about the self.[3] "Existing theoretical accounts of omniscient narration derive largely from the study of classic nineteenth century novels," Dawson points out. "While narrative theory acknowledges historical shifts in fashion, it operates with a synchronic understanding of omniscient narration as a static element of narrative, produced by the structural relationship between focalization and voice."[4] In other words, critics analyzing or theorizing the representation of the self should be aware that the notion of the self on which such representations are built changed over time. J. Paul Hunter has shown, for example, that narrative structures addressing "epistemological concerns" could be more recognizable to eighteenth-century readers than to "modern readers."[5] More complexly, sometimes literary techniques for representing the self were developed during periods in which notions of the self themselves were in flux. No wonder if representations of the self lumped in the category "literary omniscience" or "omniscient narrator" fail to fit theory's models.[6]

Scholars have long acknowledged that contextual factors significantly affect the production of text, whether those factors
are technological, economic, educational, political, social, religious, or geographical, to name only a few. When it comes to the
influence of the revolution in natural philosophy, scholars have
already begin demonstrating how artists of other forms adjusted
their techniques to reflect ideas and methods of the revolution.
Svetlana Alpers outlines the impact of natural philosophy on the
subjects and techniques of Dutch painting. By the seventeenth
century, "Links between art and the attempts of the new experimental technology to control nature were well established in the
Netherlands," she explains.[7] The impact on the arts was equally
significant in England. Marjorie Hope Nicholson describes changes
to visual imagery in eighteenth-century poetry due to the influence
of Newtonian optics. Mordechai Feingold describes English visual
culture's different techniques in response to Newtonian physics'
epistemological and cultural domination during the same period.
Laura Baudot shows how Joseph Wright of Derby fused experimental philosophy and paintings' *vanitas* tradition to create a new
system of representation. Lorraine Daston and Peter Galison track
the development of visual representations of nature in atlases in
response to epistemological changes beginning in the seventeenth
century.[8] The list could go on and on, with the inclusion of narrative structures for creating point of view.

It may be that point of view in the novel can be something of a
canary in the coal mine: significant changes to it may indicate significant changes to notions of the self. That is not to overshadow
the genius of individual authors, of course. Hobbesian mechanism
like biographical determinism is too reductive to explain fully human
creativity. Jane Austen's development of free indirect discourse in the
later stages of her career is still the product of brilliance.[9] It may be
brilliance responding in its own innovative way to ideas about the self
in circulation, however. The variety of current notions about the early
nineteenth century's ideas of the self, including a vast body of scholarship on Romanticism—especially the Romantic subject or subjectivity, Dror Wahrman's claims about the self at the end of the eighteenth
century, and Daston and Galison's very different views of the self at
the same period—suggest intriguing possibilities.

For this study, however, such possibilities must be viewed from
afar. In 1727, the novel was just getting established as a genre. After
a few decades of innovation and disrepute, it was about to ebb for a
decade before surging powerfully onto the literary scene again with

the work of authors including Samuel Richardson, Henry Fielding, Charlotte Lennox, and a returned Eliza Haywood. As we continue the project of assessing their debts to predecessors like Behn and Davys, it is worth keeping in mind the question of the relationship between structure and epistemology.

Notes

Introduction

1. Charles Taylor, *Sources of the Self: The Making of the Modern Identity* (Cambridge, MA: Harvard University Press, 1989); Timothy J. Reiss, *Mirages of the Selfe: Patterns of Personhood in Ancient and Early Modern Europe* (Stanford: Stanford University Press, 2003); Dror Wahrman, *The Making of the Modern Self: Identity and Culture in Eighteenth-Century England* (New Haven: Yale University Press, 2006). See also Roy Porter, *Flesh in the Age of Reason* (New York: W. W. Norton, 2003).
2. Amos Funkenstein, *Theology and the Scientific Imagination from the Middle Ages to the Seventeenth Century* (Princeton: Princeton University Press, 1986).
3. Wahrman, *Making of the Modern Self*; Reiss, *Mirages of the Selfe*; Taylor, *Sources of the Self*; Eve Keller, "Producing Petty Gods: Margaret Cavendish's Critique of Experimental Science," *ELH* 64, no. 2 (1997): 457, doi:10.1353/elh.1997.0017; Lorraine Daston and Peter Galison, *Objectivity*, paperback ed. (New York: Zone Books, 2010).
4. Wahrman, *Making of the Modern Self*, xii.
5. See, for example, Steven Shapin and Simon Shaffer, *Leviathan and the Air Pump: Hobbes, Boyle, and the Experimental Life. Including and Translation of Thomas Hobbes, "Dialogus Physicus de Natura Aeris," by Simon Shaffer* (Princeton: Princeton University Press, 1985); Steven Shapin, *A Social History of Truth* (Chicago: University of Chicago Press, 1994); Steven Shapin, *The Scientific Revolution* (Chicago: University of Chicago Press, 1996); Michael Hunter, *Science and the Shape of Orthodoxy: Intellectual Change in Late Seventeenth-Century Britain* (Woodbridge, UK: Boydell, 1995); Peter Dear, *Discipline and Experience: The Mathematical Way in the Scientific Revolution* (Chicago: University of Chicago Press, 1995); Judith P. Zinsser, ed., *Men, Women, and the Birthing of Modern Science* (Dekalb, IL: Northern Illinois University Press, 2005); Lynette Hunter and Sarah Hutton, Introduction to *Women, Science and Medicine 1500–1700: Mothers and Sisters of the Royal Society*, ed. Lynette Hunter and Sarah Hutton (Stroud: Sutton Publishing, 1997), 1–6.
6. See, for example, Shapin and Shaffer, *Leviathan and the Air Pump*; Shapin, *Social History*; Hunter, *Science and the Shape of Orthodoxy*; Dear, *Discipline and Experience*; Peter Harrison, "Newtonian Science, Miracles, and the Laws of Nature," *Journal of the History of Ideas* 56,

no. 4 (October 1995): 531–53, http://www.jstor.org/stable/2709991; Peter Dear, "Totius in Verba: Rhetoric and Authority in the Early Royal Society," *Isis* 76, no. 2 (June 1985): 149–61; Geoffrey Gorham, "Mind-Body Dualism and the Harvey-Descartes Controversy," *Journal of the History of Ideas* 55, no. 2 (April 1994): 211–34, doi:10.2307/2709897; Barbara M. Benedict, "The Mad Scientist: The Creation of a Literary Stereotype," in *Imagining the Sciences: Expressions of New Knowledge in the "Long" Eighteenth Century*, ed. Robert C. Leitz, III and Kevin L. Cope (New York: AMS Press, 2004), 68–70; Elizabeth Potter, *Gender and Boyle's Law of Gases* (Bloomington: Indiana University Press, 2001); Anna Battigelli, *Margaret Cavendish and the Exiles of the Mind* (Lexington, KY: The University Press of Kentucky, 1998), 107; Keller, "Producing Petty Gods"; Sasha Handley, *Visions of an Unseen World: Ghost Beliefs and Ghost Stories in Eighteenth-Century England* (London: Pickering & Chatto, 2007); Francis Young, *English Catholics and the Supernatural, 1553–1929* (Burlington: Ashgate, 2013).

7. Michael Hunter, "The Making of Christopher Wren," in Hunter, *Science and the Shape of Orthodoxy*, 49.

8. This narrative of self and epistemology follows the opposite course traced by Loraine Daston and Peter Galison in *Objectivity*. They argue that scientific methodology created a sense of self, particularly a scientific self. I do not dispute that epistemology may have generated a scientific self—one investigated not only by Daston and Galison but also, for example, by Jan Golinski in his work on Humphy Davy—but I do suggest that the epistemology whose influence they are tracking itself had its origins in something beyond itself. Daston and Galison, *Objectivity*; Jan Golinski, *Science as Public Culture: Chemistry and Enlightenment in Britain, 1760–1820* (Cambridge: Cambridge University Press, 1992).

9. Battigelli, *Margaret Cavendish*, 107. See also Keller, "Producing Petty Gods."

10. Susan Sniader Lanser, *Fictions of Authority: Women Writers and Narrative Voice* (Ithaca, NY: Cornell University Press, 1992), 6.

11. George S. Rousseau, *Enlightenment Borders: Pre- And Post-Modern Discourses: Medical, Scientific* (Manchester, UK: Manchester University Press, 1991), 284, 294; Marjorie Hope Nicolson, *Newton Demands the Muse: Newton's "Opticks" and the Eighteenth-Century Poets* (Princeton: Princeton University Press, 1946). See also Rachel Carnell, *Partisan Politics, Narrative Realism, and the Rise of the British Novel* (New York: Palgrave Macmillan, 2006), 18; Mordechai Feingold, *The Newtonian Moment: Isaac Newton and the Making of Modern Culture* (New York: The New York Public Library and Oxford University Press, 2004); Jayne Elizabeth Lewis, *Air's Appearance: Literary Atmosphere in British Fiction, 1660–1794* (Chicago: University of Chicago Press, 2012).

12. Michael Mascuch, *Origins of the Individualist Self: Autobiography and Self-Identity in England, 1591–1791* (Stanford: Stanford University Press, 1996).

13. Carnell, *Partisan Politics*; Jesse Molesworth, *Chance and the Eighteenth-Century Novel: Realism, Probability, Magic* (New York: Cambridge University Press, 2010); Ros Ballaster, *Seductive Forms: Women's Amatory Fiction from 1684 to 1740* (Oxford: Clarendon Press, 1998).

14. Londa Schiebinger, *The Mind Has No Sex? Women in the Origins of Modern Science* (Cambridge, MA: Harvard University Press, 1989); Erica Harth, *Cartesian Women: Versions and Subversions of Rational Discourse in the Old Regime* (Ithaca, NY: Cornell University Press, 1992). See also Lynette Hunter and Sarah Hutton, ed., *Women, Science and Medicine 1500–1700: Mothers and Sisters of the Royal Society* (Stroud: Sutton Publishing, 1997); Zinsser, *Men, Women, and the Birthing of Modern Science*; Judy A. Hayden, ed., *The New Science and Women's Literary Discourse: Prefiguring Frankenstein*, (New York: Palgrave Macmillan, 2011).

15. Schiebinger, *The Mind*, 41, 44; Feingold, *Newtonian Moment*, 119, 134–35; Nicolson, *Newton Demands the Muse*, 16; Rousseau, *Enlightenment Borders*, 281.

16. Schiebinger, *The Mind*, 37.

17. See, for example, Schiebinger, *The Mind*, 44–45; Lynette Hunter, "Sisters of the Royal Society: The Circle of Katherine Jones, Lady Ranelagh," in Hunter and Hutton, *Women, Science and Medicine*, 178–97; Margaret P. Hannay, "'How I These Studies Prize': The Countess of Pembroke and Elizabethan Science," in Hunter and Hutton, *Women, Science and Medicine*, 108–21.

18. Zinsser, Introduction to Zinsser, *Men, Women, and the Birthing of Modern Science*, 3–9; Hunter and Hutton, Introduction, 5; Schiebinger, *The Mind*, 10–47; Feingold, *Newtonian Moment*, 122–25; Harth, *Cartesian Women*, 5; Patricia Sheridan, Introduction to *Catharine Trotter Cockburn: Philosophical Writings*, ed. Patricia Sheridan (Toronto: Broadview, 2006), 9–11.

19. Judy A. Hayden, "Women, Education, and the Margins of Science," in Hayden, *The New Science*, 2.

20. Feingold, *Newtonian Moment*, 125.

21. Potter, *Gender and Boyle's Law*, 4.

22. Rousseau, *Enlightenment Borders*, 276–81; Kristine Larsen, "'A Woman's Place Is in the Dome': Gender and the Astronomical Observatory, 1670–1970," *MP: An Online Feminist Journal* 2, no. 5 (October 2009): 104–24, http://academinist.org/wp-content/uploads/2009/10/Woman_Place_Larsen.pdf; Feingold, *Newtonian Moment*, 119–25.

23. See chapter 3 for a more extensive discussion of the role of community in the development of epistemology and of the sense of self during this period.

24. "The Royal Society: Statistics," The Royal Society, http://royalsociety. org/about-us/equality/statistics/; Schiebinger, *The Mind*, 26.

25. Donna Haraway, "Modest_Witness@Second_Millennium," in *The Haraway Reader* (New York: Routledge, 2004), 232.

26. Frances Harris, "Living in the Neighbourhood of Science: Mary Evelyn, Margaret Cavendish and the Greshamites," in Hunter and Hutton, *Women, Science and Medicine*, 210–11; Hunter, "Sisters," 188–91.

27. Linda Alcoff and Elizabeth Potter, "Introduction: When Feminisms Intersect Epistemology," in *Feminist Epistemologies*, ed. Linda Alcoff and Elizabeth Potter (New York: Routledge, 1993), 11. See also Vrinda Dalmiya and Linda Alcoff, "Are 'Old Wives' Tales' Justified?" in Alcoff and Potter, *Feminist Epistemologies*, 217–44; Stephen Clucas, "Joanna Stephens's Medicine and the Experimental Philosophy," in Zinsser, *Men, Women, and the Birthing of Modern Science*, 142. Michael Hunter observes that starting in the 1640s, followers of the "new science" evinced a general "proneness to downgrade other forms of knowledge and other methods of ratiocination by comparison" ("Christopher Wren," 48).

28. See, for example, Golinski, *Science as Public Culture*, 3; Dear, "Totius in Verba"; Jan V. Golinski, "Robert Boyle: Skepticism and Authority in Seventeenth-Century Chemical Discourse," in *The Figural and the Literal: Problems of Language in the History of Science and Philosophy, 1630–1800*, ed. Andrew E. Benjamin, Geoffrey N. Cantor, and John R. R. Christie (Manchester, UK: Manchester University Press, 1987), 58–82; John T. Harwood, "Rhetoric and Graphics in *Micrographia*," in *Robert Hooke: New Studies*, ed. Michael Hunter and Simon Schaffer (Woodbridge, UK: Boydell, 1989), 119–47; Dwight Atkinson, *Scientific Discourse in Sociohistorical Context: "The Philosophical Transactions of the Royal Society of London," 1675–1975* (Mahwah, NJ: Lawrence Erlbaum Associates, 1999).

29. Dear, "Totius in Verba," 161.

30. As Lanser puts it in *The Narrative Act*, "If we can acknowledge the crucial impact of point of view in the presentation and reception of a tale, then surely we can also entertain the possibility of more complex links between ideology and technique through the structuring of point of view. It is possible that the very choice of a narrative technique can reveal and embody ideology." Susan Sniader Lanser, *The Narrative Act: Point of View in Prose Fiction* (Princeton: Princeton University Press, 1981), 18. See also Lanser, *Fictions of Authority*.

31. Golinski, "Robert Boyle"; Shapin and Shaffer, *Leviathan and the Air Pump*, 60–69.

32. A smattering of studies on this subject includes Jane Spencer, *The Rise of the Woman Novelist: From Aphra Behn to Jane Austen* (Oxford: Blackwell, 1986); Nancy Armstrong, *Desire and Domestic Fiction: A Political History of the Novel* (New York: Oxford University Press,

1987); Janet Todd, *The Sign of Angellica: Women, Writing, and Fiction,
1600–1800* (New York: Columbia University Press, 1989); Catherine
Gallagher, *Nobody's Story: The Vanishing Acts of Women Writers
in the Marketplace, 1670–1820* (Berkeley: University of California
Press, 1994); Josephine Donovan, *Women and the Rise of the Novel,
1405–1726* (New York: St. Martin's Press, 1999); Paula Backscheider,
ed., *Revising Women: Eighteenth-Century Women's Fiction and Social
Engagement* (Baltimore: The Johns Hopkins University Press, 2000);
Sarah Prescott, *Women, Authorship and Literary Culture, 1690–1740*
(New York: Palgrave Macmillan, 2003); Susan Carlile, ed., *Masters of
the Marketplace: British Women Novelists of the 1750s* (Bethlehem:
Lehigh University Press, 2011).

33. J. Paul Hunter, "Robert Boyle and the Epistemology of the Novel,"
Eighteenth-Century Fiction 2, no. 4 (1990): 275–91; Michael McKeon,
The Origins of the English Novel, 1660–1740 (Baltimore: The Johns
Hopkins University Press, 1987); Rebecca Tierney-Hynes, *Novel Minds:
Philosophers and Romance Readers, 1680–1740* (New York: Palgrave
Macmillan, 2012). See also Martha A. Turner, *Mechanism and the Novel:
Science in the Narrative Process* (Cambridge: Cambridge University
Press, 1993); Helen Thompson and Natania Meeker, "Empiricism,
Substance, Narrative: An Introduction," special issue of *Eighteenth
Century: Theory and Interpretation* 48, no. 3 (Fall 2007): 183–86.

34. J. A. Downie, "Mary Davys's 'Probable Feign'd Stories' and Critical
Shibboleths about 'The Rise of the Novel,'" *Eighteenth-Century Fiction*
12, no. 2–3 (January–April 2000): 309–26, doi:10.1353/ecf.2000.0033.
There are any number of studies that similarly argue that the novel
emerged from a melting pot of genres and discourses—McKeon, *Origins
of the English Novel*; Lennard Davis, *Factual Fictions* (New York:
Columbia University Press, 1983); J. Paul Hunter, *Before Novels* (New
York: W. W. Norton, 1990); William B. Warner, *Licensing Entertainment*
(Berkeley: University of California Press, 1998); Brean Hammond and
Shaun Regan, *Making the Novel: Fiction and Society in Britain, 1660–
1789* (New York: Palgrave Macmillan, 2006); Mary Poovey, *Genres of
the Credit Economy* (Chicago: University of Chicago Press, 2008); and
so forth. Downie argues convincingly not only for novelists' awareness
of this process at the time but also for the termination of a first stage or
arc of development at 1730, when novelists fell temporarily silent and
playwrights—some of them the same people—took voice.

35. Davys's novel *The False Friend* was published in 1732 but it is reprint of
The Cousins, published in 1725.

36. John Bender, *Imagining the Penitentiary: Fiction and the Architecture of
Mind in Eighteenth-Century England* (Chicago: University of Chicago
Press, 1987); Paula R. Backscheider, "The Story of Eliza Haywood's
Novels," in *The Passionate Fictions of Eliza Haywood: Essays on
Her Life and Work*, ed. Kirsten T. Saxton and Rebecca P. Bocchicchio

(Lexington, KY: The University Press of Kentucky, 2000), 25–26; Downie, "'Probable Feign'd Stories.'"

37. Carnell, *Partisan Politics*, 3.

38. Ibid., 4.

39. See, for example, Gérard Genette, *Figures of Literary Discourse*, trans. Alan Sheridan (New York: Columbia University Press, 1982); Gérard Genette, *Narrative Discourse: An Essay in Method*, trans. Jane E. Lewin (Ithaca, NY: Cornell University Press, 1980); Jonathan Culler, "Omniscience," *Narrative* 12, no. 1 (January 2004): 22–34; Nicholas Royle, *The Uncanny* (New York: Routledge, 2003), 256–63.

40. Paul Dawson, "The Return of Omniscience in Contemporary Fiction." *Narrative* 17, no. 2 (May 2009): 144, doi:10.1353/nar.0.0023.

41. Dawson, "Return," 148–50.

42. Michel Foucault, *The Archaeology of Knowledge & The Discourse on Language*, trans. A. M. Sheridan Smith (New York: Pantheon, 1972); Judith Butler, *Gender Trouble: Feminism and the Subversion of Identity* (New York: Routledge, 1990); Judith Butler, *Bodies That Matter: On the Discursive Limits of 'Sex'* (New York: Routledge, 1993); Ellen Spolsky, "Narrative as Nourishment," in *Toward a Cognitive Theory of Narrative Acts*, ed. Frederick Luis Aldama (Austin: University of Texas Press, 2010), 37–60; Lanser, *Fictions of Authority*; Lanser, *The Narrative Act*.

43. Dawson, "Return," 150.

44. Ibid., 149.

45. Blakey Vermeule writes, for example, "Rather than seeing the science of moral psychology as an effect of Enlightenment ideology, I would argue that it picks out deep facts about the sources of morality, facts that the evolutionary synthesis is now able to explain and contextualize. On this view, empiricism turns out to have the right account of the sources of moral life. This view informs the readings in this book in ways that I will explain." Blakey Vermeule, *The Party of Humanity: Writing Moral Psychology in Eighteenth-Century Britain* (Baltimore: The Johns Hopkins University Press, 2000). See also, for example, Lisa Zunshine, *Why We Read Fiction* (Columbus: The Ohio State University Press, 2006); Alan Richardson, *The Neural Sublime: Cognitive Theories and Romantic Texts* (Baltimore: The Johns Hopkins University Press, 2010); David Herman, "Narrative Theory after the Second Cognitive Revolution," in *Introduction to Cognitive Cultural Studies*, ed. Lisa Zunshine (Baltimore: The Johns Hopkins University Press, 2010), 155–75; Jonathan Kramnick, *Actions and Objects from Hobbes to Richardson* (Stanford: Stanford University Press, 2010); Patrick Colm Hogan, *The Mind and Its Stories: Narrative Universals and Human Emotion* (Cambridge: Cambridge University Press, 2003); Mark Blackwell, ed. *The Secret Life of Things: Animals, Objects, and It-Narratives in Eighteenth-Century England* (Lewisburg: Bucknell University Press, 2007); "Empiricism, Substance, Narrative," ed. Helen

Thompson and Natania Meeker, special issue, *Eighteenth-Century: Theory and Interpretation* 48, no. 3 (Fall 2007).

1 Notions of the Self

1. Timothy J. Reiss, *Mirages of the Selfe: Patterns of Personhood in Ancient and Early Modern Europe* (Stanford: Stanford University Press, 2003); Charles Taylor, *Sources of the Self: The Making of the Modern Identity* (Cambridge, MA: Harvard University Press, 1989); Amos Funkenstein, *Theology and the Scientific Imagination from the Middle Ages to the Seventeenth Century* (Princeton: Princeton University Press, 1986); and Roy Porter, *Flesh in the Age of Reason* (New York: W. W. Norton, 2003).

2. John E. Leary, Jr., *Francis Bacon and the Politics of Science* (Ames: Iowa State University Press, 1994), 146.

3. Just how long depends on which scholars one consults. Dror Wahrman regards the development of the modern self as complete by the turn of the nineteenth century; Lorraine Daston and Peter Galison see it as just beginning at that point. Dror Wahrman, *The Making of the Modern Self: Identity and Culture in Eighteenth-Century England* (New Haven: Yale University Press, 2004); Lorraine Daston and Peter Galison, *Objectivity*, paperback ed. (New York: Zone Books, 2010).

4. Francis Bacon, "The New Organon," in *The Complete Essays of Francis Bacon*, ed. Henry LeRoy Finch (New York: Washington Square Press, 1963), 195–96.

5. See, for example, Porter, *Flesh in the Age of Reason*, 24–26, 54–63.

6. Thomas Willis, *An essay of the pathology of the brain and nervous Stock in Which Convulsive Diseases are Treated of*, trans. Samuel Pordage (London: J. B., 1681).

7. Peter Dear, *Discipline and Experience: The Mathematical Way in the Scientific Revolution* (Chicago: University of Chicago Press, 1995), 6.

8. Misty G. Anderson, "Tactile Places: Materializing Desire in Margaret Cavendish and Jane Barker," *Textual Practice* 13, no. 2 (1999): 329–30.

9. Robert Boyle, *The Christian Virtuoso I*, in *The Works of Robert Boyle*, ed. Michael Hunter and Edward B. Davis, vol. 11, *"The Christian Virtuoso" and other publications of 1687–91* (London: Pickering & Chatto, 2000), 295.

10. Boyle, *The Christian Virtuoso I*, 11: 292.

11. Robert Hooke, Preface to *Micrographia* (London: Jo. Martyn and Ja. Allestry, 1665), http://www.gutenberg.org/files/15491/15491-h/15491-h.htm.

12. See also Porter, *Flesh in the Age of Reason*; Eve Keller, "Producing Petty Gods: Margaret Cavendish's Critique of Experimental Science," *ELH* 64, no. 2 (1997): 457, doi:10.1353/elh.1997.0017. Wahrman dates the

primary phase in the reorganization of ideas of self too late, after the middle of the eighteenth century. Nevertheless, his larger point that "the supposed universality of the individual subject with a well-defined, stable, unique, centered self is in truth a charged, far from natural, recent Western creation" remains valid. Wahrman, *The Making of the Modern Self*, xii. This self ought not to be confused with the objective self discussed at great length by Lorraine Daston and Peter Galison in *Objectivity*. While Daston and Galison's characterization of seventeenth-century epistemology and what they call science is unfortunately total-izing given the debates of that period, their definitions of objectivity and the practices it required and created are indeed different from those of seventeenth-century and early eighteenth-century natural philosophers.

13. Reiss, *Mirages of the Selfe*, 4.
14. Under "Gender," the Index to Steven Shapin's *The Social History of Truth* says, "See Women," as if women have gender but men do not. Steven Shapin and Simon Shaffer, *Leviathan and the Air Pump: Hobbes, Boyle, and the Experimental Life. Including and Translation of Thomas Hobbes, "Dialogus Physicus de Natura Aeris," by Simon Shaffer* (Princeton: Princeton University Press, 1985); Steven Shapin, *A Social History of Truth* (Chicago: University of Chicago Press, 1994), 473; Philip Carter, *Men and the Emergence of Polite Society, Britain, 1660–1800* (Harlow: Pearson, 2001).
15. Donna Haraway, "Modest_Witness@Second_Millennium," in *The Haraway Reader* (New York: Routledge, 2004), 225.
16. Elizabeth Potter, *Gender and Boyle's Law of Gases* (Bloomington: Indiana University Press, 2001), 3.
17. Potter, *Gender and Boyle's Law*; Haraway, "Modest Witness"; Rachel Carnell, *Partisan Politics, Narrative Realism, and the Rise of the British Novel* (New York: Palgrave Macmillan, 2006), 17–23.
18. Shapin, *Social History*, 405.
19. Keller, "Producing Petty Gods," 451.
20. Carole Pateman, *The Sexual Contract* (Stanford: Stanford University Press, 1988). See also Kathryn Shevelow, *Women and Print Culture: The Construction of Femininity in the Early Periodical* (New York: Routledge: 1989), 12–13; Carnell, *Partisan Politics*; Rachel Weil, *Political Passions: Gender, the Family, and Political Argument in England, 1680–1714* (Manchester, UK: Manchester University Press, 1999); Elizabeth Fox Genovese, *Feminism Without Illusions: A Critique of Individualism* (Chapel Hill: The University of North Carolina Press, 1991), 16–17. This absorption lingers on: Galen Strawson explains that "a human being (a *man*, in Locke's terminology, in which 'man' refers equally to male and female)" without considering the ramifications of this maneuver by either himself or by Locke. Galen Strawson, *Locke on Personal Identity: Consciousness and Concernment* (Princeton: Princeton University Press, 2011), 7, original italics.

21. Erica Harth, *Cartesian Women: Versions and Subversions of Rational Discourse in the Old Regime* (Ithaca, NY: Cornell University Press, 1992), 9.
22. Joan Wallach Scott, *Gender and the Politics of History*, rev. ed. (New York: Columbia University Press, 1999), 25, eBook.
23. Scott, *Gender*, 25.
24. Potter, *Gender and Boyle's Law*. See also Elizabeth Grosz, "Bodies and Knowledges: Feminism and the Crisis of Reason," in Alcoff and Potter, *Feminist Epistemologies*, 187–215.
25. Laura Brown, "Feminization of Ideology: Form and the Female in the Long Eighteenth Century," in *Ideology and Form in Eighteenth-Century Literature*, ed. David H. Richter (Lubbock: Texas Tech University Press, 1999), 226, 231.
26. Thomas Laqueur, *Making Sex: Body and Gender from the Greeks to Freud* (Cambridge, MA: Harvard University Press, 1990); Karen Harvey, *Reading Sex in the Eighteenth Century: Bodies and Gender in English Erotic Culture* (Cambridge: Cambridge University Press, 2004); Karen Harvey, "The Substance of Sexual Difference: Change and Persistence in Representations of the Body in Eighteenth-Century England," *Gender & History* 14, no. 2 (August 2002): 202–23; Laura Gowing, *Common Bodies: Women, Touch and Power in Seventeenth-Century England* (New Haven: Yale University Press, 2003).
27. Isaac Newton, "Preface to the Reader" in *Principia Naturalis*, trans. J. Bruce Brackenridge, in Brackenridge, *Key to Newton's Dynamics: The Kepler Problem and the Principia: Containing an English Translation of Sections 1, 2, and 3 of Book One From the First (1687) Edition of Newton's Mathematical Principles of Natural Philosophy* (Berkeley: University of California Press, 1995), 230, eBook.
28. Newton, Preface, 231.
29. Dear, *Discipline*, 2–3, 210–48. See also Peter Dear, *The Intelligibility of Nature: How Science Makes Sense of the World* (Chicago: University of Chicago Press, 2006), 26–31.
30. Judith P. Zinsser, "The Many Representations of the Marquise Du Châtelet," in *Men, Women, and the Birthing of Modern Science*, ed. Judith P. Zinsser (Dekalb, IL: Northern Illinois University Press, 2005), 55.
31. John Locke, *An Essay Concerning Human Understanding*, 6th ed. (Kitchener, ON: Batoche Books, 2001), II.xxvii.9, eBook, original italics.
32. For more discussion of Locke and Newton's recognition of the affinity between their ideas and method see, for example, Peter Walmsley, *Locke's "Essay" and the Rhetoric of Science* (Lewisburg: Bucknell University Press, 2003), 20; Richard S. Westfall, *Never at Rest: A Biography of Isaac Newton*, paperback reprint ed. (Cambridge: Cambridge University Press, 1984), 488–93; Michael Hunter, "The Debate over Science," in *Science and the Shape of Orthodoxy: Intellectual Change in Late*

Seventeenth-Century Britain (Woodbridge: Boydell Press, 1995), 119; Mordechai Feingold, *The Newtonian Moment: Isaac Newton and the Making of Modern Culture* (New York: The New York Public Library and Oxford University Press, 2004), 44; Robert Iliffe, *Newton: A Very Short Introduction* (New York: Oxford University Press, 2007), 35, 103–4, eBook.

33. Edmund Halley, "On this Work of Mathematical Physics Of the Most Outstanding Man Most Learned Isaac Newton," in Newton, *Principia Naturalis*, 233.

34. Thomas Burnet, *Remarks upon an Essay concerning Humane Understanding, in a Letter to the Author* (London: M. Wotton, 1697), 5.

35. See, for example, Dear, *Discipline and Experience*; Michael Hunter, "The Making of Christopher Wren," in *Science and the Shape of Orthodoxy*, 55.

36. William Petty, *Another Essay in Political Arithmetick, Concerning the Growth of the City of London with the Measures, Periods, Causes, and Consequences Thereof, 1682* (London: H. H. for Mark Pardoe, 1683), np.

37. William Petty, "An Extract of Two Essays in Political Arithmetick concerning the Comparative Magnitudes, etc. of London and Paris by Sr. William Petty Knight. R. S. S.," *Philosophical Transactions* 16 (1686–92): 152, http://www.jstor.org/stable/101856.

38. The Preface to *Philosophical Transactions*, 16 (1686–92): 36, http://www.jstor.org/stable/101838.

39. See, for example, John W. Yolton, *A Locke Dictionary* (Cambridge, MA: Blackwell, 1993), 53; J. Joanna S. Forstrom, *John Locke and Personal Identity: Immortality and Bodily Resurrection in 17th-Century Philosophy* (New York: Continuum, 2010); Walmsley, *Locke's "Essay"*; Porter, *Flesh in the Age of Reason*, 55–77; Edwin McCann, "Locke's Philosophy of Body," in *The Cambridge Companion to Locke*, ed. Vere Chappell (New York: Cambridge University Press, 1994), 56–60; Steven Shapin, *Never Pure: Historical Studies of Science as if It was produced by People with Bodies, Situated in Time, Space, Culture, and Society, and Struggling for Credibility and Authority* (Baltimore: The Johns Hopkins University Press, 2010), 169–70; Patricia Sheridan, Introduction to *Catharine Trotter Cockburn: Philosophical Writings*, ed. Patricia Sheridan (Toronto: Broadview, 2006), 17–19.

40. Locke, *Essay*, 13.

41. Walmsley, *Locke's "Essay,"* 17.

42. For a more thorough review of the debates over the nature of the self as it involved a soul see, for example, Porter, *Flesh in the Age of Reason*, 60–76; Christopher Fox, *Locke and the Scriblerians: Identity and Consciousness in Early Eighteenth-Century Britain* (Berkeley: University of California Press, 1988), 14–17; McCann, "Locke's Philosophy of Body," 76–86. See also Funkenstein on theology's replacement by natural philosophy and the idea of the scientist as "secular theologian" in

Theology and the Scientific Imagination. A sampling of texts from the 1720s that rejected Locke's theories include Henry Felton's *Resurrection of the same numerical body* (Oxford, 1725), Philips Gretton's *A review of the argument a priori, In Relation to the Being and Attributes of God* (London: Bernard Lintot, 1726), and Richard Greene's *The Principles of the Philosophy of the Expansive and Contractive Principles* (Cambridge: Cornelius Crownfield, 1727).

43. George S. Rousseau, *Enlightenment Borders: Pre-And Post-Modern Discourses: Medical, Scientific* (Manchester, UK: Manchester University Press, 1991), 269; Barbara Shapiro, "History and Natural History in Sixteenth- and Seventeenth-Century England: An Essay on the Relationship between Humanism and Science," in *English Scientific Virtuosi in the 16th and 17th Centuries*, papers read at a Clark Library Seminar, February 3, 1977, ed. Barbara Shapiro and Robert G. Frank (Los Angeles: William Andrews Clark Memorial Library, 1977), 321; Michael Hunter, "Introduction: Fifteen Essays and a New Theory of Intellectual Change," in *Science and the Shape of Orthodoxy,* 6; I. Bernard Cohen and Richard S. Westfall, General Introduction in *Newton: Texts, Backgrounds, Commentaries,* ed. I. Bernard Cohen and Richard S. Westfall (New York: W. W. Norton, 1995), xiv. Shapiro asserts that it was the *Principia,* not the *Opticks* that had such a significant impact, but she was speaking of the impact on natural philosophy and I am speaking of the impact on culture.

44. Marjorie Hope Nicolson, *Newton Demands the Muse: Newton's "Opticks" and the Eighteenth-Century Poets* (Princeton: Princeton University Press, 1946), 1–14; Feingold, *The Newtonian Moment,* 169–90.

45. Winch Holdsworth, *A Sermon Preached Before the University of Oxford at St Mary's on Easter-Monday....* (Oxford: 1920); Catherine Trotter Cockburn, *A Letter to Dr. Holdsworth, Occasioned by His Sermon Preached Before the University of Oxford....* (London: Benjamin Motte, 1726); Joseph Butler, *The Analogy of Religion, Natural and Revealed....*(Dublin: J. Jones, 1736); Vincent Perronet, *A Vindication of Mr. Locke* (London: James, John, and Paul Knapton, 1736).

46. Keller, "Producing Petty Gods," 457.

47. Reiss, *Mirages of the Selfe,* 3.

48. Bacon, "The New Organon," 193–95, 189.

49. René Descartes, *Discourse on Method,* in *Discourse on Method and Meditations on First Philosophy,* trans. Donald A. Cress, 4th ed. (Indianapolis: Hackett, 1998), 8, 13.

50. See, for example, Peter Harrison, "Curiosity, Forbidden Knowledge, and the Reformation in Natural Philosophy in Early Modern England," *Isis* 92, no. 2 (June 2001): 265–90, http://www.jstor.org/stable/3080629; Michael McKeon, *The Origins of the English Novel, 1660–1740* (Baltimore: The Johns Hopkins University Press, 1987); Brian Worden, "The Question of Secularization," in *A Nation Transformed: England*

after the Restoration, ed. Alan Houston and Steve Pincus (Cambridge: Cambridge University Press, 2001), 20–40; Barbara Shapiro, "Natural Philosophy and Political Periodization: Interregnum, Restoration and Revolution," in Houston and Pincus, *A Nation Transformed*, 303; Shapiro, "History and Natural History," 3–55; Laura Baudot, "An Air of History: Joseph Wright's and Robert Boyle's Air Pump Narratives," *Eighteenth-Century Studies* 46, no. 1 (Fall 2012): 1–28.

51. Reiss, *Mirages of the Selfe*, 2. See also Taylor, *Sources of the Self*, 139–40; Peter Dear, "From Truth to Disinterestedness in the Seventeenth Century," *Social Studies of Science* 22, no. 4 (November 1992), 628, http://www.jstor.org/stable/285457.

52. Daston and Galison, *Objectivity*, 203.

53. Dear, "From Truth to Disinterestedness," 627; Harrison, "Curiosity." Dear, writing well before Lorraine Daston and Peter Galison's book *Objectivity*, uses "objectivity" and "disinterestedness" interchangeably, a looseness of terminology perhaps less possible in the wake of Daston and Galison's book. As a result, I speak of "detachment" but not of "objectivity" in this study.

54. Boyle, *The Christian Virtuoso I*, 285, 293.

55. Jane Barker, "An Invitation to my Friends at Cambridge," in *Poetical Recreations: Consisting of Original Poems, Songs, Odes etc; With several New Translations. In Two Parts: Part I. Occasionally written by Mrs. Jane Barker. Part II. By Several Gentlemen of the Universities, and Others* (London: Benjamin Crayle, 1688). A revised version of this poem appeared in *A Patch-Work Screen for the Ladies* (1723). For more discussion of Jane Barker and natural philosophy, please see Chapter 3.

56. Shapin, *Never Pure*, 119–23. See also Steven Shapin, *A Social History of Truth* (Chicago: University of Chicago Press, 1994).

57. Lorraine Code, "Taking Subjectivity into Account," in *Feminist Epistemologies*, ed. Linda Alcoff and Elizabeth Potter (New York: Routledge, 1993), 17.

58. Dear, *Discipline and Experience*, 210, 216, 232, 247; Shapiro, "Natural Philosophy," 319–20. Dwight Atkinson notes, for example, that in 1725 the reports in *Philosophical Transactions* were more mathematically based than in the late seventeenth century. Dwight Atkinson, *Scientific Discourse in Sociohistorical Context: "The Philosophical Transactions of the Royal Society of London," 1675–1975* (Mahwah, NJ: Lawrence Erlbaum Associates, 1999), 77, 85, 145–47.

59. Jesse Molesworth, *Chance and the Eighteenth-Century Novel: Realism, Probability, Magic* (Cambridge: Cambridge University Press, 2010), 9.

60. Helene Moglin, although dating the realist novel to much later in the century than most scholars and certainly than Molesworth or Carnell, contends that realism's interior self excluded any notion of female self-hood or female experience. Helene Moglin, *The Trauma of Gender: A*

Feminist Theory of the English Novel (Berkeley: University of California Press, 2001); Carnell, *Partisan Politics*.

61. Shapin and Shaffer, *Leviathan and the Air Pump*, 25.
62. William Harvey, *On the Motion of the Heat and Blood in Animals*, trans. Robert Willis (Buffalo: Prometheus Books, 1993), 9.
63. Shapin, *Social History*, 7.
64. Colin Maclaurin, *An Account of Sir Isaac Newton's Philosophical Discoveries, in Four Books....*(London: A. Millar and J. Nourse, 1748), 68.
65. See, for example, Michael Honeybone, "Sociability, Utility and Curiosity in the Spalding Gentlemen's Society, 1710–60," in *Science and Beliefs: From Natural Philosophy to Natural Science, 1700–1900*, ed. David M. Knight and Matty D. Eddy (Aldershot: Ashgate, 2005), 63–76; Peter Clark, *British Clubs and Societies, 1500–1800* (New York: Oxford University Press, 2000).
66. Shapin, *Social History*, 7; Shapin and Shaffer, *Leviathan and the Air Pump*, 55, 25.
67. Frederick Slare, *Experiments and Observations upon Oriental and Other Bezoar-Stones, Which Prove Them to be of No Use in Physic....* (London: Timothy Goodwin, 1715), i–ii.
68. Shapin presents this disagreement between Boylean experimentalists and Newtonian mathematicians in implicitly democratic terms: "circulation of knowledge in public space was deemed vital for securing its veracity and legitimacy," he observes. The definitions of "public space" or "public forums of truth-making" are, however, not about the larger society but about those already in philosophical circles. Shapin, *Social History*, 335–37.
69. Robert Boyle, *Some Considerations Touching the Usefulness of Experimental Natural Philosophy. The First Part*, in *The Works of Robert Boyle*, ed. Michael Hunter and Edward B. Davis, vol. 3, *"The Usefulness of Natural Philosophy" and sequels to "The Spring of the Air," 1662–3* (London: Pickering & Chatto, 1999), 199.
70. Michael Hunter, "The Debate over Science," in *Science and the Shape of Orthodoxy*, 106–7; Harrison, "Curiosity," 274, 278; Shapiro, "Natural Philosophy," 307.
71. Hooke, Preface to *Micrographia*, np.
72. Thomas Morgan, *Philosophical Principles of Medicine....* (London: J. Darby and T. Browne, 1725), v.
73. Aaron Hill, *An Account of the Rise and Progress of the Beech-Oil Invention, and All the Steps Which Have Been Taken in That Affair, from the First Discovery, to the Present Time. As Also, What is further design'd in That Undertaking. By Aaron Hill Esq.* (London: [s.n.], 1715).
74. Christine Gerrard, *Aaron Hill: The Muses' Projector, 1685–1750* (Oxford: Oxford University Press, 2003), 39–40.

75. Eliza Haywood, *The Female Spectator*, in *Selections from "The Female Spectator*," ed. Patricia Meyers Spacks (New York: Oxford University Press, 1999), III, XV, 195.

76. Reiss, *Mirages of the Selfe*, 3.

77. Harrison, "Curiosity," 287–88.

78. Karen Harvey, "Refinement in a Teacup? Punch, Domesticity and Gender in the Eighteenth Century," *Journal of Design History* 21, no. 3 (Autumn 2008): 205–21, doi:10.1093/jdh/epn022; Terry Lovell, "Subjective Powers?: Consumption, the Reading Public, and Domestic Woman in Early Eighteenth-Century England," in *The Consumption of Culture*, ed. Ann Bermingham and John Brewer (New York: Routledge, 1995), 34–35.

79. Richard Allestree, Preface to *The Ladies Calling* (Oxford: Printed at the Theatre, 1673), np.

80. See, for example, Will Pritchard, *Outward Appearances: The Female Exterior in Restoration London* (Lewisburg: Bucknell University Press, 2008).

81. For extensive analyses of how gender and a gendered epistemology were constructed by experimental philosophers during the late seventeenth and early eighteenth centuries, see, for example, Haraway, "Modest_ Witness," 223–50; Potter, *Gender and Boyle's Law*; Vrinda Dalmiya and Linda Alcoff, "Are 'Old Wives' Tales' Justified?" in Alcoff and Potter, *Feminist Epistemologies*, 217–44.

82. Michelle M. Dowd and Julie A. Eckerle, Introduction to *Genre and Women's Life Writing in Early Modern England*, ed. Michelle M. Dowd and Julie A. Eckerle (Burlington: Ashgate, 2007), 4. See also Jo Wallwork and Paul Salzman, ed., *Early Modern Englishwomen Testing Ideas* (Burlington: Ashgate, 2011); Michael Mascuch, *Origins of the Individualist Self* (Stanford: Stanford University Press, 1996).

83. Diana B. Altegoer, *Reckoning Words: Baconian Science and the Construction of Truth in English Renaissance Culture* (Madison, NJ: Fairleigh Dickinson University Press, 2000). See also Walmsley, *Locke's "Essay"*; Hannah Dawson, *Locke, Language and Early-Modern Philosophy* (Cambridge: Cambridge University Press, 2007); Jan Golinski, "Robert Boyle: Skepticism and Authority in Seventeenth-Century Chemical Discourse," in *The Figural and the Literal: Problems of Language in the History of Science and Philosophy, 1630–1800*, ed. Andrew E. Benjamin, Geoffrey N. Cantor, and John R. R. Christie (Manchester, UK: Manchester University Press, 1987), 64–65; John T. Harwood, "Rhetoric and Graphics in *Micrographia*," in *Robert Hooke: New Studies*, ed. Michael Hunter and Simon Schaffer (Woodbridge: Boydell Press, 1989), 119–47; Sarah Hutton, "The Riddle of the Sphinx: Francis Bacon and the Emblems of Science," in *Women, Science and Medicine 1500–1700: Mothers and Sisters of the Royal Society*, ed. Lynette Hunter and Sarah Hutton (Stroud: Sutton Publishing, 1997), 7–28; Funkenstein, *Theology and the Scientific Imagination*, 28–29.

84. Michael Hunter, *Boyle: Between God and Science* (New Haven: Yale University Press, 2009), 93–95; Leary, *Francis Bacon*, ix; Jo Wallwork, "Disruptive Behaviour in the Making of Science: Cavendish and the Community of Seventeenth-Century Science," in Wallwork and Salzman, *Early Modern Englishwomen*, 49–50. Other examples of contemporary texts concerned with the reformation of language, particularly to support "the new learning," include Francis Lodowyck, *The Ground-Work, or Foundation, Laid (or so Intended) for the Framing of a New Perfect Language....* (London: [s.n], 1652); Henry Edmundson, *Lingua Linguarum....* (London: T. Roycroft, 1655); R. F., *The Pure Language of the Spirit of Truth....* (London: Giles Calvert, 1655); or John Webb, *An Historical Essay Endeavoring a Probability that the Language of the Empire of China is the Primitive Language* (London: Nathaniel Brook, 1669).

85. Altegoer, *Reckoning Words*, 23.

86. Shapin and Shaffer, *Leviathan and the Air Pump*, 65.

87. See also Anne Bratach, "Following the Intrigue: Aphra Behn, Genre, and Restoration Science," *Journal of Narrative Technique* 26, no. 3 (Fall 1996): 213.

88. See, for example, Jane Donawerth, "Conversation and the Boundaries of Public Discourse in Rhetorical Theory by Renaissance Women," *Rhetorica: A Journal of the History of Rhetoric* 16, no. 2 (Spring 1998): 181–99, http://www.jstor.org/stable/10.1525/rh.1998.16.2.181; Wallwork, "Disruptive Behaviour," 41–53; Ros Ballaster, "Taking Liberties: Revisiting Behn's Libertinism," *Women's Writing* 19, no. 2 (May 2012): 165–76, http://dx.doi.org/10.1080/09699082.2011.646861.

89. Keller, "Producing Petty Gods."

90. Shapin and Shaffer, *Leviathan and the Air Pump*, 66. See also Altegoer, *Reckoning Words*, 24.

91. Potter, *Gender and Boyle's Law*, 10.

92. Amy Elizabeth Smith, "Naming the Un-'Familiar': Formal Letters and Travel Narratives in Late Seventeenth- and Eighteenth-Century Britain," *The Review of English Studies* 54, no. 214 (May 2003): 180. For more discussion of epistolary subgenres and their use, particularly in seventeenth-century natural philosophy, see for example Diana Barnes, "Familiar Epistolary Philosophy: Margaret Cavendish's *Philosophical Letters* (1664)," *Parergon* 26, no. 2 (2009): 39–64; Diana Barnes, "The Restoration of Royalist Form in Margaret Cavendish's *Sociable Letters*," in *Women Writing 1550–1750*, ed. Jo Wallwork and Paul Salzman, special issue of *Meridian: The La Trobe University English Review* 18, no. 1 (2001): 201–214; Anne L. Bower, "Dear –: In Search of New (Old) Forms of Critical Address," in *Epistolary Histories*, ed. Amanda Gilroy and W. M. Verhoeven (Charlottesville: University Press of Virginia, 2000),155–75.

93. Barnes, "The Restoration of Royalist Form," 205.

94. Gerald MacLean, "Re-Siting the Subject," in Gilroy and Verhoeven, *Epistolary Histories*, 177, 182.

95. Cavendish's *Sociable Letters* (1664) and *Philosophical Letters* (1664) both used epistolary form, for example. Boyle wrote letters for *Philosophical Transactions,* and he also used epistolary form in other publications, such as *New Experiments Physico-Mechanical* (1662) and *The Aerial Noctiluca* (1680).

96. Boyle, *The Christian Virtuoso I,* 11: 291.

97. Atkinson, *Scientific Discourse,* 81. Because the letters were often excerpted according to their title in *Philosophical Transactions,* presumably Atkinson depends on the letters' self-identifying as epistles for these statistics.

98. Roger Iliffe, "Author-Mongering: The 'Editor' Between Producer and Consumer," in Bermingham and Brewer, *The Consumption of Culture 1600–1800,* 173, original italics.

99. Anonymous, "A Narrative Concerning the Success of Pendulum-Watches at Sea for the Longitudes," *Philosophical Transactions* 1, no. 1 (1665–66): 13–14; http://www.jstor.org/stable/101409.

100. P. L., *Two Essays Sent in a Letter from Oxford, to a Nobleman in London....* (London: R. Baldwin, 1695).

101. Harvey, *On the Motion,* 32.

102. Harwood, "Rhetoric and Graphics," 138.

103. Hooke, *Micrographia,* np.

104. Atkinson, *Scientific Discourse,* 75–77.

105. Anonymous, "A Narrative Concerning the Success of Pendulum-Watches at Sea for the Longitudes," 13, 14. "Major Holmes" is something of a shady character, evidently. In addition to his contributions to natural philosophy, this Major Holmes appears to have abetted Lord Argyll's escape from prison in 1681 and supported the Rye House Plot in 1683. *Dictionary of National Biography* (1886; Hathi Trust Digital Library), ed. Leslie Stephen, vol. 8 (London: Smith, Elder, 1886), 336–37.

106. Antoine van Leeuwenhoek, "Microscopical Observations of Mr. Leewenhoeck, Concerning the Optic Nerve, Communicated to the Publisher in Dutch, and by Him Made English," part 1, *Philosophical Transactions* 10 (1675): 379, http://www.jstor.org/stable/101663.

2   An Ingenious Romance: The Stable Self

1. Roger Cotes, "Cotes' Preface to the Second Edition," in Isaac Newton, *The Mathematical Principles of Natural Philosophy and His System of the World,* trans. Andrew Motte (1729), ed. Florian Cajoli (1934), vol. 1: *The Motion of Bodies* (New York: Greenwood Press, 1969), xx.

2. Recently, Leah Orr has challenged the attribution of *Love-Letters.* In the absence of more consistent logic and other scholarship, I have

chosen to follow the established practice of scholars in the field and treat *Love-Letters* as Behn's. Even if *Love-Letters* is not Behn's, this chapter shows that its handling of the self is consistent with Behn's interest in natural philosophy expressed in her narrative fiction, poetry, drama, and translations, and supports my larger argument that authors of the period found the idea of the stable, unified self profoundly and increasingly problematic as political events unfolded. Leah Orr, "Attribution Problems in the Fiction of Aphra Behn," *The Modern Language Review* 108, no. 1 (January 2013): 30–51, http://www.jstor.org/stable/10.5699/modelangrevi.108.1.0030.

3. A sampling of scholarship examining Behn's familiarity with natural philosophy includes Anne Bratach, "Following the Intrigue: Aphra Behn, Genre, and Restoration Science," *Journal of Narrative Technique* 26, no. 3 (Fall 1996): 209–227; Sarah Goodfellow, "'Such Masculine Strokes': Aphra Behn as Translator of *A Discovery of New Worlds*," *Albion: A Quarterly Journal Concerned with British Studies* 28, no. 2 (Summer 1996): 229–250; Line Cottegnies, "The Translator as Critic: Aphra Behn's Translation of Fontenelle's *Discovery of New Worlds* (1688)," *Restoration* 27, no. 1 (Spring 2003): 23–38; Helen Thompson, "'Thou Monarch of my Panting Soul': Hobbesian Obligation and the Durability of Romance in Aphra Behn's *Love-Letters*," in *British Women's Writing in the Long Eighteenth Century: Authorship, Politics and History*, ed. Jennie Batchelor and Cora Kaplan (New York: Palgrave Macmillan, 2005),107–120; Barbara M. Benedict, "The Curious Genre: Female Inquiry in Amatory Fiction," *Studies in the Novel* 30, no. 2 (Summer 1998): 194–210; Al Coppola, "Retraining the Virtuoso's Gaze: Behn's *Emperor of the Moon*, The Royal Society, and the Spectacles of Science and Politics," *Eighteenth-Century Studies* 41, no. 4 (Summer 2008): 481–506; Alvin Snider, "Atoms and Seeds: Aphra Behn's Lucretius," *Clio* 33, no. 1 (Fall 2003): 1–24; Alvin Snider, "Cartesian Bodies," *Modern Philology: A Journal Devoted to Research in Medieval and Modern Literature* 92, no. 2 (November 2000): 299–319; Ros Ballaster, "Taking Liberties: Revisiting Behn's Libertinism," *Women's Writing* 19, no. 2 (May 2012): 165–76, http://dx.doi.org/10.1080/09699082.2011.646861; Karen Gevirtz, "Aphra Behn and the Scientific Self," in *The New Science and Women's Literary Discourse: Prefiguring Frankenstein*, ed. Judy Hayden (New York: Palgrave Macmillan, 2011), 85–98; Maureen Duffy, *The Passionate Shepherdess: Aphra Behn, 1640–89* (Jonathan Cape: London, 1977); Janet Todd, *The Secret Life of Aphra Behn* (London: Pandora, 2000); Angeline Goreau, *Reconstructing Aphra: A Social Biography of Aphra Behn* (New York: Dial, 1980).

4. Mary Ann O'Donnell traces the development of Behn's abilities as a translator in several languages in the introduction to *Aphra Behn: An Annotated Bibliography of Primary and Secondary Sources*, 2nd ed.

(Burlington: Ashgate, 2004), 4. O'Donnell suggests that while Behn's French was strong enough by the late 1680s to produce her supple translation of Fontenelle and her reflections on translating in the "Translator's Preface" to that work, Behn's Latin probably required "help" from someone more fluent. (In the textual introduction to Volume 1 of *The Works of Aphra Behn*, Todd speculates in a brief narrative about what that process might have looked like). Whatever Behn's linguistic proficiencies, it remains striking that three of her translations or associations with significant translations are works of natural philosophy when she could have chosen or been chosen to translate other kinds of texts. As Todd notes, "In her prose translations Behn appears to have followed Roscommon and Dryden in choosing a source consistent with her own views and temperament." Janet Todd, textual introduction to *The Works of Aphra Behn*, ed. Janet Todd, vol. 1, *Poetry* (Columbus: Ohio State University Press, 1992), xxxvi; Todd, textual introduction to *The Works of Aphra Behn*, ed. Janet Todd, vol. 4, *Translations* (Columbus: Ohio State University Press, 1993), ix. See also Snider, "Atoms and Seeds," 15.

5. Aphra Behn, *The Roundheads*, in *The Works of Aphra Behn*, ed. Janet Todd, vol. 6, *The Plays, 1678–1682* (Columbus: University of Ohio Press, 1996), 385. I am indebted to Mary Ann O'Donnell for pointing out this reference.

6. Aphra Behn, "To the Unknown Daphnis on his Excellent Translation of Lucretius," in *The Works of Aphra Behn*, ed. Janet Todd, vol. 1, *Poetry* (Columbus: Ohio State University Press, 1992), 7–10. For a more extensive treatment of Behn's poem, see Snider, "Atoms and Seeds," 1–24.

7. Coppola, "Retraining the Virtuoso's Gaze"; Duffy, *The Passionate Shepherdess*, 263, 270–274; Goodfellow, "'Such Masculine Strokes'"; Cottegnies, "The Translator as Critic"; Thompson, "'Thou Monarch of my Panting Soul'"; Goreau, *Reconstructing Aphra*, 141, 168, 187, 182; Todd, *Secret Life*, 290–94; Susan Staves, *A Literary History of Women's Writing in Britain, 1660–1789* (New York: Cambridge University Press, 2006), 80–81.

8. Benedict, "The Curious Genre," 194–210.

9. Aphra Behn, *Love-Letters Between a Nobleman and His Sister*, in *The Works of Aphra Behn*, ed. Janet Todd, vol. 2, *Love-Letters Between a Nobleman and His Sister* (Columbus: University of Ohio Press, 1993), 32, 158, 185–186. All further references to this text are abbreviated *Love-Letters*.

10. Bratach, "Following the Intrigue," 211–213, 219; Aphra Behn, *Oroonoko*, ed. Joanna Lipking (New York: W. W. Norton, 1997), 8. See also Benedict, "The Curious Genre," 197.

11. George Warren, *An impartial description of Surinam upon the continent of Guiana in America* (London: William Godbid, 1667), 10–11. An entertaining specimen of this type of writing is Henry Stubbe's very

serious anthropology of chocolate, *The Indian nectar, or, A discourse concerning chocolate....*(London: J. C., 1662).

12. Henry Power, *Experimental Philosophy, In Three Books: Containing New Experiments Microscopical, Mercurial, Magnetical....* (London: T. Roycroft, 1664), np. It is worth noting that Power's book appeared before Hooke's *Micrographia*, even if it was not as sumptuous, and used many of the same descriptive and organizational techniques.

13. Maximilian E. Novak, "Friday: or, the Power of Naming," in *Augustan Subjects: Essays in Honor of Martin C. Battestin*, ed. Albert J. Rivero (Newark, DE: University of Delaware Press, 1997), 110–22.

14. Emily Hodgson Anderson, "Novelty in Novels: A Look at What's New in Aphra Behn's *Oroonoko*," *Studies in the Novel* 39, no. 1 (Spring 2007): 4.

15. Anderson, "Novelty in Novels," 5–6.

16. For more discussion of spectacle and the scientific revolution, see for example Coppola, "Virtuoso's Gaze"; Tita Chico, "Gimcrack's Legacy: Sex, Wealth, and the Theater of Experimental Philosophy," *Comparative Drama* 42, no. 2 (Spring 2008): 29–49; M. A. Katritzky, *Women, Medicine and Theatre, 1500–1750* (Burlington: Ashgate, 2007); Bratach, "Following the Intrigue," 220–21; Bernadette Bensaude-Vincent and Christine Blondel, eds., *Science and Spectacle in the European Enlightenment* (Burlington: Ashgate, 2008).

17. Behn's description of the "numb eel" resembles Warren's in his *Impartial History of Surinam*, but as Stanley Finger recently pointed out, Behn's sources must have gone well beyond Warren and probably included personal experience. Finger also points out that the properties of such sea creatures were often described as "cold." Oroonoko's complaint about the effects being inconsistent with "cold philosophy" suggests another reading, as well. Cold was a topic that fascinated seventeenth-century natural philosophers, particularly empiricists. Boyle conducted an extensive series of experiments to investigate cold as a natural phenomenon, for example, and published the results in *New Experiments and Observations Touching Cold, Or an Experimental History of Cold*. Warren, *Impartial History*, 2; Stanley Finger, "The Lady and the Eel: How Aphra Behn Introduced Europeans to the "Numb Eel," *Perspectives in Biology and Medicine* 55, no. 3 (Summer 2012): 378–401; Robert Boyle, *New Experiments and Observations Touching Cold, Or an Experimental History of Cold....*, in *The Works of Robert Boyle*, ed. Michael Hunter and Edward B. Davis, vol. 4, *Colours* and *Cold*, 1664–5 (London: Pickering & Chatto, 1999). I am indebted to Mary Ann O'Donnell for calling my attention to Finger's essay.

18. Stephen Clucas describes a series of experiments performed on a mixture developed by Joanna Stephens for curing gallstones, pointing out that one after the other, the experimenters tested the medicine on themselves and others without considering the impact of the body on the results of

the experiment. Similarly, Newton is famous for performing a number of his experiments in optics on his own eyes, and Boyle tested the powers of cold fluids by drinking them. Stephen Clucas, "Joanna Stephens's Medicine and the Experimental Philosophy," in *Men, Women, and the Birthing of Modern Science*, ed. Judith P. Zinsser (Dekalb, IL: Northern Illinois University Press, 2005), 149–50; Isaac Newton, Laboratory Notebook, c. 1669–1693, MS Add. 3975, Newton Papers, Cambridge University Library, Cambridge, UK; George Johnson, *The Ten Most Beautiful Experiments* (New York: Knopf, 2008), 38–39; Boyle, *Experimental History of Cold*, 4:343–44.

19. Alvin Snider points out that from Lucretius, Behn developed the idea that desire was a force of nature the way Newton would later identify gravity as a force of nature. For Behn, Snider shows, desire was thus a predictable, ineluctable force that acted on the body (Snider, "Atoms and Seeds," 4). For a different interpretation of the episode of the numb eel as commentary on natural philosophy, see Bratach, "Following the Intrigue," 217; Finger, "The Lady and the Eel," 388–89.

20. Thomas O. Beebee, *Epistolary Fiction in Europe: 1500–1850* (Cambridge: Cambridge University Press, 1999), 173; John Richetti, "*Love Letters Between a Nobleman and His Sister*: Aphra Behn and Amatory Fiction," in Rivero, *Augustan Subjects*, 21; Janet Todd, General Introduction to *The Works of Aphra Behn*, vol. 1, *Poetry*, xxi; Todd, *Secret Life*, 337.

21. A. Smith, "Naming the Un-'Familiar,'" 178–202.

22. Diana Barnes, "Familiar Epistolary Philosophy: Margaret Cavendish's *Philosophical Letters* (1664)," *Parergon* 26, no. 2 (2009): 39–64; Elspeth Graham, "Intersubjectivity, Intertextuality, and Form in the Self-Writings of Margaret Cavendish," in Dowd and Eckerle, *Genre and Women's Life Writing*, 131–50; Jan Golinski, "Robert Boyle: Skepticism and Authority in Seventeenth-Century Chemical Discourse," in *The Figural and the Literal: Problems of Language in the History of Science and Philosophy, 1630–1800*, ed. Andrew E. Benjamin, Geoffrey N. Cantor, and John R. R. Christie (Manchester, UK: Manchester University Press, 1987), 58–82.

23. Janet Todd, "*Love-Letters* and Critical History," in *Aphra Behn (1640–1689): Identity, Alterity, Ambiguity*, ed. Mary Ann O'Donnell, Bernard Dhuicq and Guyonne Leduc (Montreal: L'Harmattan, 2000), 198; Diana Barnes, "The Restoration of Royalist Form in Margaret Cavendish's *Sociable Letters*," in *Women Writing 1550–1750*, ed. Jo Wallwork and Paul Salzman, special issue of *Meridian: The La Trobe University English Review* 18, no. 1 (2001): 205–6.

24. Helen Wilcox, "'Free and Easy as ones discourse'?: Genre and Self-Expression in the Poems and Letters of Early Modern Englishwomen," in Dowd and Eckerle, *Genre and Women's Life Writing*, 20.

25. Gerald MacLean, Postscript to Gilroy and Verhoeven, *Epistolary Histories*, 172.

26. See, for example, Ruth Perry, *Women, Letters, and the Novel* (New York: AMS Press, 1980), 4, 14, 75; Beebee, *Epistolary Fiction*; Barbara Maria Zaczek, *Censored Sentiments: Letters and Censorship in Epistolary Novels and Conduct Material* (Newark, DE: University of Delaware Press, 1997); Susan Wright, "Private Language Made Public: The Language of Letters as Literature," *Poetics* 18 (1989): 549–78; Gary Schneider, *The Culture of Epistolarity: Vernacular Letters and Letter Writing in Early Modern England, 1500–1700* (Newark, DE: University of Delaware Press, 2005).

27. Beebee, *Epistolary History*, 118. See also, for example, Tony Davies, "The Ark in Flames: Science, Language and Education in Seventeenth-Century England," in Benjamin, Cantor, and Christie, *The Figural and the Literal*, 83–102.

28. Amanda Gilroy and W. M. Verhoeven, Introduction to Gilroy and Verhoeven, *Epistolary Histories*, 1.

29. Gilroy and Verhoeven, Introduction, 5, 1.

30. Donald R. Wehrs, "*Eros*, Ethics, and Identity: Royalist Feminism and the Politics of Desire in Aphra Behn's *Love-Letters Between a Nobleman and His Sister*," *SEL* 32 (1992): 461.

31. Behn, *Love-Letters*, 236.

32. Thomas Sprat, *History of the Royal Society*, ed. Jackson I. Cope and Harold Whitmore Jones (St. Louis: Washington University Press, 1958), 113.

33. Wehrs, "*Eros*, Ethics, and Identity," 461, 466; Maureen Duffy, Introduction to *Love-Letters Between a Nobleman and His Sister* by Aphra Behn (New York: Virago: 1987), x–xi. See also Jorge Figueroa-Dorrego, "Reconciling 'the most Contrary and Distant Thoughts': Paradox and Irony in the Novels of Aphra Behn," in *Re-Shaping the Genres: Restoration Women Writers*, ed. Zenón Luis-Martinez and Jorge Figueroa-Dorrego (New York: Peter Lange, 2003), 239.

34. Todd, General Introduction, 1:xi.

35. Margarete Rubik, "Estranging the Familiar, Familiarizing the Strange: Self and Other in *Oroonoko* and *The Widdow Ranter*," in O'Donnell, Dhuicq, and Leduc, *Aphra Behn (1640–1689)*, 40. See also Annamaria Lamarra, "The Difficulty in Saying 'I': Aphra Behn and the Female Autobiography," in *Aphra Behn (1640–1689): Le Modèle Européen*, ed. Mary Ann O'Donnell and Bernard Dhuicq (Entrevaux, France: Bilingua GA Editions, 2005), 1–7.

36. Helen Thompson also argues that *Love-Letters* is critical of the role of the body and of language in politics, but her perspective is somewhat different than mine. Thompson suggests that Behn indicts both Hobbesian mechanism through the body, which supports monarchy by the unsexed abjuration of individualistic desire, and arbitrary signification in the form of words, which threaten monarchy because they can be used for flattery and to deceive. According to Thompson, both systems

exist at equal levels of approval and disapproval in the novel. Thompson, "'Thou Monarch of my Panting Soul.'"

37. For more on the problems of the physically absent beloved in epistolary fiction, see Stephen Ahern, "'Glorious ruine': Romantic Excess and the Politics of Sensibility in Aphra Behn's *Love-Letters*," *Restoration* 29, no. 1 (Spring 2005): 30.

38. Ballaster, "Taking Liberties"; Snider, "Atoms and Seeds."

39. Snider, "Atoms and Seeds,"19.

40. Snider, "Atoms and Seeds."

41. Rubik, "Estranging the Familiar," 33.

42. Todd, *Secret Life*, 299–302.

43. See, for example, Peter Walmsley, *Locke's "Essay" and the Rhetoric of Science* (Lewisburg: Bucknell University Press, 2003).

44. See also Ros Ballaster, *Seductive Forms: Women's Amatory Fiction from 1684 to 1740* (Oxford: Clarendon Press, 1998), 11. Ballaster argues that the late seventeenth-century and early eighteenth-century political conflicts were still open to women, who could participate in them through the prose fictions, especially the amatory narratives, that became the novel.

45. See, for example, John E. Leary, Jr., *Francis Bacon and the Politics of Science* (Ames: Iowa State University Press, 1994); James Dougal Fleming, "Introduction: The Invention of Discovery, 1500–1700," in *The Invention of Discovery, 1500–1700*, ed. James Dougal Fleming (Burlington: Ashgate, 2011), 8.

46. Peter Laslett, Introduction to *Two Treatises of Government* by John Locke, ed. Peter Laslett, student edition (Cambridge: Cambridge University Press, 1988), 51; Rachel Carnell, *Partisan Politics, Narrative Realism, and the Rise of the British Novel* (New York: Palgrave Macmillan, 2006), 20, 4–8. See also Gary S. De Krey, "Radicals, Reformers and Republicans: Academic Language and Political Discourse in Restoration London," in *A Nation Transformed: England after the Restoration*, ed. Alan Houston and Steve Pincus (Cambridge: Cambridge University Press, 2001), 71–99; Francis F. Steen, "The Politics of Love: Propaganda and Structural Learning in Aphra Behn's Love-Letters between a Nobleman and His Sister," *Poetics Today* 23, no. 1 (Spring 2002): 91–122; Jacqueline Broad, "Mary Astell's Machiavellian Moment? Politics and Feminism in *Moderation truly Stated*," in *Early Modern Englishwomen Testing Ideas*, ed. Jo Wallwork and Paul Salzman (Burlington: Ashgate, 2011), 9–23.

47. Carnell, *Partisan Politics*, 10; Ballaster, *Seductive Forms*, 11, 78–79.

48. "Like Fielding, Behn knows that historical events can be distorted in historical narratives," he points out. Albert J. Rivero, "'Heiroglyphick'd' History: in Aphra Behn's *Love-Letters between a Nobleman and His Sister*," *Studies in the Novel* 30, no. 2 (Summer 1998): 134.

49. The *Oxford English Dictionary* offers one military definition for "advantage": "A favourable place for defence or attack, *esp.* an elevated

place, a vantage point." The latest date offered by the OED for this usage is 1666, suggesting not only that Behn's narrative may have been one of its last appearances but also that she had an interesting command of military terminology. *Oxford English Dictionary*, s.v. "advantage," accessed November 5, 2012, http://www.oed.com.ezproxy.shu.edu/sear ch?searchType=dictionary&q=advantage&_searchBtn=Search.

50. Although at present there is insufficient archival material to date precisely many of Behn's short fictional narratives, it is generally accepted that they are products of the 1680s. Without more information, it is only safe to say that Behn's short prose fiction confirms Behn's ongoing interest over the 1680s in the self, its capacities, and how it is or should be represented in writing.

51. Aphra Behn, "The Unfortunate Happy Lady," in *The Works of Aphra Behn*, ed. Janet Todd, Vol. 3: *"The Fair Jilt" and Other Stories* (Columbus: Ohio State University Press, 1995), 382.

52. Aphra Behn, "The Adventure of the Black Lady," in Todd, *The Works of Aphra Behn*, 3:320, 3:319.

53. Aphra Behn, "The History of the Nun," in Todd, *The Works of Aphra Behn*, 3:212–13.

54. Mary Ann O'Donnell considers this passage potentially autobiographical, "although the narrator's voice may be that of a *persona* that ought not be identified as Behn's." Janet Todd also suggests that there are autobiographical elements in some of Behn's short fiction, not just in *Oroonoko*. This passage certainly bears a resemblance to a passage in *Love-Letters* in which the narrator describes her ecstasy during church services. Without further evidence, however, such speculations must remain exactly that. O'Donnell, *Annotated Bibliography*, 2; Janet Todd, Textual introduction to *The Works of Aphra Behn*, ed. Janet Todd, 3:ix; Behn, *Love-Letters*, 397.

55. Rubik, "Estranging the Familiar," 33–41.

56. For more discussion of the use of the body in Harvey or Hooke's writing, please see chapter 1.

57. Rupert Hall, "Isaac Newton: Creator of the Cambridge Scientific Tradition," in *Cambridge Scientific Minds*, ed. Peter Harman and Simon Mitton (Cambridge: Cambridge University Press, 2002), 40–41.

58. Isaac Newton, *Sir Isaac Newton's Mathematical Principles of Natural Philosophy and His System of the World*, trans. Andrew Motte (1729), ed. Florian Cajoli (1934), vol. 1: *The Motion of Bodies* (New York: Greenwood Press, 1962), 5, 4.

59. It was established practice during the late seventeenth century to use the particular as the general. The maneuver, according to Peter Dear, came from Aristotelian methodology and had not yet been objected to. "Above all," Dear notes, "throughout the century the universal experience reigned virtually unchallenged as the irreducible touchstone of empirical adequacy." Peter Dear, *Discipline and Experience: The Mathematical Way in the Scientific Revolution* (Chicago: University of Chicago Press, 1995), 6.

60. Hooke is quite explicit in the Preface of *Micrographia*, in fact: "If therefore the Reader expects any infallible Deductions, or certainty of Axioms, I am to say of my self, that those stronger Works of Wit and Imagination are above my weak Abilities; or if they had not been so, I would not have made use of them in this present Subject before me...." Robert Hooke, Preface to *Micrographia* (London: Jo. Martyn and Ja. Allestry, 1665), http://www.gutenberg.org/files/15491/15491-h/15491-h.htm.

61. Steven Shapin and Simon Shaffer, *Leviathan and the Air Pump: Hobbes, Boyle, and the Experimental Life. Including and Translation of Thomas Hobbes, "Dialogus Physicus de Natura Aeris," by Simon Shaffer* (Princeton: Princeton University Press, 1985), 60–69.

62. I am indebted to Cynthia Richards for this point.

63. Robert L. Chibka suggests that the cultural differences between Oroonoko and the English, including the narrator, are to no small degree based in the willingness to manipulate truth. "Europeans continually maintain power over Oroonoko by a twofold mechanism," Chibka writes; "they lie and assume that he does the same. He, by the same token, remains powerless because he tells truth and assumes that they will do the same." Robert L. Chibka, "'Oh! Do Not Fear a Woman's Invention': Truth, Falsehood, and Fiction in Aphra Behn's *Oroonoko*," *Texas Studies in Literature and Language* 30, no. 4 (Winter 1998): 520.

64. Rubik, "Estranging the Familiar," 33.

65. Jacqueline Pearson, "Gender and Narrative in the Fiction of Aphra Behn," part 1, *The Review of English Studies* 42, no. 165 (February 1991): 43, original italics.

3 The Fly's Eye: The Composite Self

1. Iain Pears, *An Instance of the Fingerpost* (New York: Berkley Books, 1998), 561.

2. Steven Shapin and Simon Shaffer, *Leviathan and the Air Pump: Hobbes, Boyle, and the Experimental Life. Including and Translation of Thomas Hobbes, "Dialogus Physicus de Natura Aeris," by Simon Shaffer* (Princeton: Princeton University Press, 1985), 15, 19.

3. Joseph Glanvill, *Scepsis Scientifica: Or, Confest Ignorance, the way to Science; In an Essay of The Vanity of Dogmatizing, and Confident Opinion, with a Reply to the Exceptions of the Learned Thomas Albius* (London: E. Cotes, 1665), np.

4. Shapin and Shaffer, *Leviathan and the Air Pump*, 23–25; Barbara Shapiro, "History and Natural History in Sixteenth- and Seventeenth-Century England: An Essay on the Relationship between Humanism and Science," in *English Scientific Virtuosi in the 16th and 17th Centuries*, papers read at a Clark Library Seminar, February 3, 1977, ed. Barbara

Shapiro and Robert G. Frank, Jr. (Los Angeles: William Andrews Clark Memorial Library, 1977), 23–30.

5. Shapin and Shaffer, *Leviathan and the Air Pump*, 55–59.

6. Donna Haraway, "Modest_Witness@Second_Millennium," in *The Haraway Reader* (New York: Routledge, 2004), 224.

7. Peter Dear, "Totius in Verba: Rhetoric and Authority in the Early Royal Society," *Isis* 76, no. 2 (June 1985): 145–56.

8. Barbara Shapiro, "Natural Philosophy and Political Periodization: Interregnum, Restoration and Revolution," in *A Nation Transformed: England after the Restoration*, ed. Alan Houston and Steve Pincus (Cambridge: Cambridge University Press, 2001), 319–20; Robert Iliffe, "Author-Mongering: The 'Editor' between Producer and Consumer," in *The Consumption of Culture 1600–1800: Image, Object, Text*, ed. Ann Bermingham and John Brewer (New York: Routledge, 1995), 167–78; Peter Dear, *Discipline and Experience: The Mathematical Way in the Scientific Revolution* (Chicago: University of Chicago Press, 1995), 2–4, 210–29; Dear, "Totius in Verba"; Steven Shapin, *Never Pure: Historical Studies of Science as if It Was produced by People with Bodies, Situated in Time, Space, Culture, and Society, and Struggling for Credibility and Authority* (Baltimore: The Johns Hopkins University Press, 2010), 133–34; Steven Shapin, *A Social History of Truth* (Chicago: University of Chicago Press, 1994), 312, 336–38; Jan Golinski, *Science as Public Culture: Chemistry and Enlightenment in Britain, 1760–1820* (Cambridge: Cambridge University Press, 1992), 5–6.

9. Toni Bowers, *Force or Fraud: British Seduction Stories and the Problem of Resistance, 1660–1760* (Oxford: Oxford University Press, 2011), 32.

10. Bowers, *Force or Fraud*; Toni O'Shaunessy Bowers, "Sex, Lies and Invisibility: Amatory Fiction from the Restoration to Mid-Century," in *The Columbia History of the British Novel*, ed. John Richetti (New York: Columbia University Press, 1994), 52; Kathleen Lubey, "Eliza Haywood's Amatory Aesthetic," *Eighteenth-Century Studies* 39, no. 3 (2006): 309–22, doi:0.1353/ecs.2006.0010.

11. See, for example, Ian Watt, *The Rise of the Novel: Studies in Defoe, Richardson and Fielding* (Berkeley: University of California Press, 1957); Catherine Gallagher, *Nobody's Story: The Vanishing Acts of Women Writers in the Marketplace, 1670–1820* (Berkeley: University of California Press, 1994); Janet Todd, *The Sign of Angellica: Women, Writing and Fiction, 1660–1800* (New York: Columbia University Press, 1989); Michael McKeon, *The Origins of the English Novel, 1660–1740* (Baltimore: The Johns Hopkins University Press, 1987); Walter L. Reed, *An Exemplary History of the Novel* (Chicago: University of Chicago Press, 1981); William B. Warner, *Licensing Entertainment: The Elevation of Novel Reading in Britain, 1684–1750* (Berkeley: University of California Press, 1998); Susan Paterson Glover, *Engendering Legitimacy: Law, Property, and Early Eighteenth-Century Fiction* (Lewisburg: Bucknell University Press, 2006).

12. Kathryn King, *Jane Barker, Exile: A Literary Career, 1675–1725* (New York: Oxford University Press, 2000), 17.

13. King, *Jane Barker*, 18.

14. King, *Jane Barker*, 22–24, 42. For examples of scholarship situating Barker within different communities, please see Rivka Swenson, "Representing Modernity of Jane Barker's *Galesia Trilogy*: Jacobite Allegory and the Patch-Work Aesthetic," *Studies in Eighteenth-Century Culture* 34 (2005): 55–80; Tonya Moutray McArthur, "Jane Barker and the Politics of Catholic Celibacy," *SEL Studies in English Literature 1500–1900* 47, no. 3 (Summer 2007): 595–618, doi:10.1353/sel.2007.0030; Niall MacKenzie, "Jane Barker, Louise Hollandine of the Palatinate and 'Solomons Wise Daughter,'" *The Review of English Studies* New Series 58, no. 233 (2007): 64–72, doi:10.1093/res/hgl142; James Fitzmaurice, "Jane Barker and the Tree of Knowledge at Cambridge University," *Renaissance Forum* 3, no. 1 (Spring 1998): np, http://www.hull.ac.uk/renforum/v3no1/fitzmaur.htm; James Fitzmaurice, "Daring and Innocence in the Poetry of Elizabeth Rochester and Jane Barker," *In-Between: Essays & Studies in Literary Criticism* 11, no. 1 (March 2002): 25–43; Kathryn R. King, "Of Needles and Pens and Women's Work," *Tulsa Studies in Women's Literature* 14, no. 1 (Spring 1995): 77–93; Kathryn King, "Jane Barker, *Poetical Recreations*, and the Sociable Text," *ELH* 61, no 3 (Autumn 1994): 551–70, http://www.jstor.org/stable/2873334. For a discussion of morality and the isolated self, please see chapters 3 and 4 of this study.

15. Josephine Donovan, "Women and the Framed-Novelle: A Tradition of Their Own," *Signs* 22, no. 4 (Summer 1997): 957, http://www.jstor.org/stable/3175225.

16. Donovan, "Framed-Nouvelle," 955; King, *Jane Barker*, 76.

17. Donovan, "Framed-Nouvelle," 952–53.

18. Robert Hooke, "Observ. 39. Of the Eyes and Head of a Grey drone-Fly, and of several other creatures," in *Micrographia* (London: Ja. Allestree, 1664), np., http://www.gutenberg.org/files/15491/15491-h/15491-h.htm.

19. See, for example, King, *Jane Barker*; Swenson, "Representing Modernity"; Carol Shiner Wilson, introduction to *The Galesia Trilogy and Selected Manuscript Poems of Jane Barker*, ed. Carol Shiner Wilson (Oxford: Oxford University Press, 1997), xv–xliii; Jane Spencer, "Creating the Women Writer: The Autobiographical Works of Jane Barker," *Tulsa Studies in Women's Literature* 2, no. 2 (Autumn 1983): 165–81, http://www.jstor.org/stable/463718. Even Josephine Donovan's examination of frame narratives in the early novel considers the Galesia narrative in *Patch-Work Screen* to be autobiography (973).

20. King, *Jane Barker*, 7.

21. Kathryn R. King and Jesslyn Medoff, "Jane Barker and Her Life (1652–1732): The Documentary Record," *Eighteenth-Century Life*

21 (November 1997): 18–22; King, *Jane Barker*, 69–73. See also Fitzmaurice, "Tree of Knowledge."

22. King, *Jane Barker*, 94. For a discussion of Barker's medical knowledge and career, see King, *Jane Barker*, 68–97.

23. Jane Barker, "An Invitation to My Friends at Cambridge," in *Poetical Recreations: Consisting of Original Poems, Songs, Odes etc; With several New Translations. In Two Parts: Part I. Occasionally Written by Mrs. Jane Barker. Part II By Several Gentlemen of the Universities, and Others* (London: Benjamin Crayle, 1688), np.

24. Misty G. Anderson, "Tactile Places: Materializing Desire in Margaret Cavendish and Jane Barker," *Textual Practice* 13, no. 2 (1999): 329–52.

25. In her poem "On my Mother and my Lady W – who both lay sick at the same time under the Hands of Dr. Paman" from *Poetical Recreations* (1688), the speaker wishes that "May all the Blessings light on thee that can / Attend a Doctor, or a Christian Man," another instance of this association in Barker's work. Jane Barker, "On my Mother and my Lady W – who both lay sick at the same time under the Hands of Dr. Paman," in *Poetical Recreations*, np.

26. Jane Barker, "A Patch-Work Screen for the Ladies," in *The Galesia Trilogy and Selected Manuscript Poems of Jane Barker*, ed. Carol Shiner Wilson (Oxford: Oxford University Press, 1997), 82. All further references to this text are abbreviated *PWS*.

27. Jane Barker, *Love Intrigues*, in *The Galesia Trilogy and Selected Manuscript Poems of Jane Barker*, ed. Carol Shiner Wilson (Oxford: Oxford University Press, 1997), 35. All further references to this text are abbreviated *LI*. King has established that *Love Intrigues* was published in 1713 by Edmund Curll without Barker's permission and under a title selected by Curll, rather than by the author. When this novel was revised and reissued in 1719, it was sent forth with Barker's original title, *The Amours of Bosvil and Galesia*. The second edition was prepared with print publication in mind while the former was written for manuscript circulation, resulting in certain differences (King, *Jane Barker*, 182–88). I am using the 1713 edition, entitled *Love Intrigues*, because this is the one available in Carol Shiner Wilson's excellent collection, *The Galesia Trilogy and Selected Manuscript Poems of Jane Barker* and in Paula R. Backscheider and John J. Richetti's groundbreaking *Popular Fiction by Women 1660–1730: An Anthology* (Oxford: Oxford University Press, 1996), 81–111.

28. Shapin and Shaffer, *Leviathan and the Air Pump*, 60.

29. According to its second charter (1663), The Royal Society's "studies are to be applied to further promoting by the authority of experiments the sciences of natural things and of useful arts, to the glory of God the Creator, and the advantage of the human race." About useful knowledge, Bacon wrote, "I would address one general admonition to all: that they consider what are the true ends of knowledge, and that they seek it...for

the benefit and use of life" (75). Bacon also had tremendous respect for agricultural knowledge. In the Preparative to *The Great Instauration*, he devotes a number of histories to agricultural topics, such as "History of the seasons or temperatures of the year," "History of plants, trees, shrubs, herbs; and of their parts, roots, stalks, wood, leaves, flowers, fruits, seeds, gums, etc.," and histories of vegetables, fishes, birds, and quadrupeds" (200–202). In *The New Atlantis*, Bacon devotes an extensive description to animal husbandry as well as agriculture and horticulture. "We have also large and various orchards and gardens, wherein we do not so much respect beauty as variety of ground and soil" because of the plenitude of food and drink that variety can produce, the "father" of Salomon House explains to the narrator of *The New Atlantis*, and describes breeding and experimenting on "beasts and birds" as well as "serpents, worms, flies, fishes" (263–64). Charles II, Second Charter of the Royal Society for Improving Natural Knowledge (1663). http://royalsociety.org/Charters-of-the-Royal-Society/; Francis Bacon, *The New Atlantis*, in *Francis Bacon: Selected Philosophical Works*, ed. Rose-Mary Sargent (Indianapolis: Hackett, 1999), 239–68; Francis Bacon, *The Great Instauration*, in Sargent, *Francis Bacon*, 66–89.

30. Lubey, "Eliza Haywood's Amatory Aesthetic," 309–10, original italics.
31. Eliza Haywood, "The British Recluse," in Backscheider and Richetti, *Popular Fiction*, 165. All further references to this text are abbreviated *BR*.
32. Fitzmaurice, "Jane Barker and the Tree of Knowledge," np. See also Fitzmaurice, "Daring and Innocence" for a discussion of Barker's revisions to poems from *Poetic Recreations* that she included in *A Patch-Work Screen*.
33. King, *Jane Barker*, 13–15.
34. Dear, *Discipline and Experience*, 210, 216, 232, 247; Shapiro, "Natural Philosophy," 319–20. For a more extensive discussion of Newton in the 1720s, please see chapter 3 of this study.
35. See, for example, Peter Harrison, "Newtonian Science, Miracles, and the Laws of Nature," *Journal of the History of Ideas* 56, no. 4 (October 1995): 531–53, http://www.jstor.org/stable/2709991; Justin E. H. Smith, "Spirit as Intermediary in Post-Cartesian Natural Philosophy," in *Spirits Unseen: The Representation of Subtle Bodies in Early Modern European Culture*, ed. Christine Göttler and Wolfgang Neuber (Boston: Brill, 2008), 269–91; McKeon, *Origins*, 75–85; Keith Hutchinson, "What Happened to Occult Qualities in the Scientific Revolution?" *Isis* 73, no. 2 (June 1982): 233–53, http://www.jstor.org/stable/231676.
36. Amos Funkenstein, *Theology and the Scientific Imagination from the Middle Ages to the Seventeenth Century* (Princeton: Princeton University Press, 1986). See also Michael Hunter, *Boyle: Between God and Science* (New Haven: Yale University Press, 2009), 200–202.
37. Peter Harrison, *The Fall of Man and the Foundations of Science* (New York: Cambridge University Press, 2007); John Henry, "Religion and

the Scientific Revolution," in *The Cambridge Companion to Science and Religion*, ed. Peter Harrison (New York: Cambridge University Press, 2010), 39–58; Francis Young, *English Catholics and the Supernatural, 1553–1929* (Burlington: Ashgate, 2013), 19–20, 25, 2, 66.

38. Jane Barker, *A Lining for the Patch-Work Screen for the Ladies*, in Wilson, *The Galesia Trilogy*, 177. All further references to this text are abbreviated *LPWS*.

39. I follow Francis Young in using "supernatural" in the sense of "all spiritual powers—good, evil or neutral" for thinking during this period. Young observes that "beliefs in the immanence of spiritual power in the world, whether for good or evil, tend to correlate. Likewise, in the early modern period, scepticism concerning miracles tended to go hand-in-hand with scepticism concerning demonic activity." Young, *English Catholics*, 4–5.

40. Jennifer Frangos, "Ghosts in the Machine: The Apparition of Mrs. Veal, Rowe's *Friendship in Death* and the Early Eighteenth-Century Invisible World," in Göttler and Neuber, *Spirits Unseen*, 315, 317.

41. Sasha Handley, *Visions of an Unseen World: Ghost Beliefs and Ghost Stories in Eighteenth-Century England* (London: Pickering & Chatto, 2007), 32.

42. Misty Anderson, for example, has argued that "Barker's organicist world emphasizes a mutual relationship between knowledge and nature which rejects both the terms of empirical objectivity and a Baconian mastery of male scientist over feminized nature" ("Tactile Places," 341).

43. Handley, *Visions*, 47. See also Young, *English Catholics*; Frangos, "Ghosts in the Machine"; Christine Göttler, Preface: "Vapours and Veils: The Edge of the Unseen," in Göttler and Neuber, *Spirits Unseen*, xv–xxvii; and J. Smith, "Spirit as Intermediary," 269–91. Barbara M. Benedict discusses immateriality in eighteenth-century discourse in "The Spirit of Things," in *The Secret Life of Things: Animals, Objects, and It-Narratives in Eighteenth-Century England*, ed. Mark Blackwell (Lewisburg: Bucknell University Press, 2007), 19–42. King also notes that the characters in Barker's reminiscences of St. Germain "seem themselves almost ghostly as they flicker in and out of the loosely connected tales, recalling a world that exists only in memory of imagination" (*Jane Barker*, 165).

44. Troy Boone, "Narrating the Apparition: Glanvill, Defoe, and the Rise of Gothic Fiction," *The Eighteenth Century* 35, no. 2 (1994): 173.

45. J. Smith, "Spirit as Intermediary," 276–77, 283–85; Henry, "Religion and the Scientific Revolution," 48–49; John W. Yolton, *Thinking Matter: Materialism in Eighteenth-Century Britain* (Minneapolis: University of Minnesota Press, 1983).

46. J. Smith, "Spirit as Intermediary," 283.

47. Handley, *Visions*, 32.

48. Aurelio M. Espinosa has identified the major permutations of this tale between the thirteenth and seventeenth centuries in France and Spain,

pointing out that versions of the story were readily available during this period. Barker's version borrows the premise of the story, but in addition, it is dominated by devices common to medieval versions, such as the initial episode in which a husband kills a lover (although this being a narrative by Barker rather than Haywood or Behn, he is only a would-be lover), the would-be lover is a member of the clergy, the corpse is placed on horseback, and the horse chases after another horse. Aurelio M. Espinosa, "Hispanic Versions of the Tale of the Corpse Many Times 'Killed,'" *The Journal of American Folklore* 49, no. 193 (July–September 1936): 182–83, http://www.jstor.org/stable/535399. Since Barker spent a large portion of her life in France, it is not surprising to find her borrowing from the French fabliau tradition. I am indebted to Angela Weisl for introducing me to this source for Barker's narrative.

49. McKeon, *Origins*, 80.

50. Leigh A. Eicke, "Jane Barker's Jacobite Writings," in *Women's Writing and the Circulation of Ideas: Manuscript Publication in England, 1550–1800*, ed. George L. Justice and Nathan Tinker (Cambridge: Cambridge University Press, 2002), 150.

51. McKeon, *Origins*, 66; Peter Harrison, *The Bible, Protestantism, and the Rise of Natural Science* (New York: Cambridge University Press, 1998); Funkenstein, *Theology and the Scientific Method*, 49.

52. Reudiger Heinze, "Violations of Mimetic Epistemology in First-Person Narrative Fiction," *Narrative* 16, no. 3 (October 2008): 280, doi:10.1353/nar.0.0008.

53. Gérard Genette, *Figures of Literary Discourse*, trans. Alan Sheridan (New York: Columbia University Press, 1982), 138–42.

54. As Carol Shiner Wilson points out, Barker adapts a number of Aphra Behn's narratives in *The Lining of the Patch-Work Screen*, including *The History of the Nun*, which becomes "Philinda's Story out of the Book" (214–17) and *The Wandering Beauty*, which becomes "The History of The Lady Gypsie" and "The History of Tangerine" (227–37) (Wilson, 214n1, 227n1).

55. Elizabeth Potter, "Gender and Epistemic Negotiation," in *Feminist Epistemologies*, ed. Linda Alcoff and Elizabeth Potter (New York: Routledge, 1993), 161.

56. Lynn Hankinson Nelson, "Epistemological Communities," in Alcoff and Potter, *Feminist Epistemologies*, 123, 124, original italics.

57. Swenson, "Representing Modernity," 58.

4   The Detached Observer

1. Isaac Newton, Laboratory Notebook, c. 1669–1693, Newton Papers, MS Add.3975, Cambridge University Library. For a very readable

description of Newton's experiment, see George Johnson, *The Ten Most Beautiful Experiments* (New York: Knopf, 2008), 38–39. Briefer accounts can be found in Richard S. Westfall, *Never at Rest: A Biography of Isaac Newton*, paperback reprint edition (Cambridge: Cambridge University Press, 1984), 94 and James Gleick, *Isaac Newton* (New York: Vintage Books, 2003), 61. A more sensational and fictionalized "account" appears in the play *Isaac's Eye* by Lucas Hnath (dir. Linsay Firman, perf. Jeff Biehl, Kristen Bush, Haskell King, and Michael Louis Serafin-Wells, Ensemble Studio Theatre, New York, 2013).

2. A detailed account of Newton's increasing sociability and use of social networks appears in Westfall's magisterial biography of Newton. As Westfall succinctly puts it, after 1689, "Newton also began to perceive himself in a new light which was incompatible with the isolation he had striven to maintain for twenty years" (481). See also Robert Iliffe, "Author-Mongering: The 'Editor' between Producer and Consumer," in *The Consumption of Culture 1600–1800: Image, Object, Text*, ed. Ann Bermingham and John Brewer (New York: Routledge, 1995), 173–78; Robert Iliffe, *Newton: A Very Short Introduction* (New York: Oxford University Press, 2007); Rupert Hall, "Isaac Newton: Creator of the Cambridge Scientific Tradition," in *Cambridge Scientific Minds*, ed. Peter Harman and Simon Mitton (Cambridge: Cambridge University Press, 2002), 46–47; or Gale E. Christianson, *Isaac Newton* (New York: Oxford University Press, 2005).

3. Gleick, *Isaac Newton*, 3.

4. Barbara M. Benedict, "The Curious Genre: Female Inquiry in Amatory Fiction," *Studies in the Novel* 30, no. 2 (Summer 1998): 194–210.

5. Joseph Drury, "Haywood's Thinking Machines," *Eighteenth-Century Fiction* 21, no. 2 (Winter 2008–9): 201–4.

6. See, for example, Paula R. Backscheider, "The Story of Eliza Haywood's Novels," in *The Passionate Fictions of Eliza Haywood: Essays on Her Life and Work*, ed. Kirsten T. Saxton and Rebecca P. Bocchicchio (Lexington, KY: The University Press of Kentucky, 2000), 31–36; Eve Tavor Bannet, "Haywood's Spectator and the Female World," in *Fair Philosopher: Eliza Haywood and "The Female Spectator,"* ed. Lynn Marie Wright and Donald J. Newman (Lewisburg: Bucknell University Press, 2006), 97–101; Kathleen Lubey, "Eliza Haywood's Amatory Aesthetic," *Eighteenth-Century Studies* 39, no. 3 (2006): 309–22; Juliette Merritt, *Beyond Spectacle: Eliza Haywood's Female Spectators* (Toronto: University of Toronto Press, 2006), 9; Earla Wilputte, Introduction to *Three Novellas* by Eliza Haywood, ed. Earla Wilputte (East Lansing: Colleagues Press, 1995), 8, 12; Aleksondra Hultquist, "Marriage in Haywood; or, Amatory Reading Rewarded," in *Masters of the Marketplace: British Women Novelists of the 1750s*, ed. Susan Carlile (Bethlehem: Lehigh University Press, 2011), 31–46; Karen Cajka,

"The Unprotected Woman in Eliza Haywood's *The History of Jemmy and Jenny Jessamy*," in Carlile, *Masters*, 47–58. For more discussion of Aphra Behn's treatment of the body in constructing the self, please see chapter 2.

7. See, for example, Ann Messenger, *His and Hers: Essays in Restoration and Eighteenth-Century Literature* (Lexington, KY: The University Press of Kentucky, 1986), 127, 133–35; Paula R. Backscheider, "The Shadow of an Author: Eliza Haywood," *Eighteenth-Century Fiction*, 11, no. 1 (October 1998): 79, http://digitalcommons.mcmaster.ca/ecf/vol11/iss1/1; Bannet, "Haywood's Spectator," 98; Kristen M. Girten, "Unsexed Souls: Natural Philosophy as Transformation in Eliza Haywood's Female Spectator," *Eighteenth-Century Studies* 43, no. 1 (Fall 2009): 55–74.

8. Christine Gerrard, *Aaron Hill: The Muses' Projector, 1685–1750* (Oxford: Oxford University Press, 2003), 9–10, 29–54.

9. Backscheider, "Shadow," 79–102.

10. Drury, "Haywood's Thinking Machines"; Tiffany Potter, "'A God-like Sublimity of Passion': Eliza Haywood's Libertine Consistency," *The Eighteenth-Century Novel* 1 (2001): 95–126.

11. Louis-Adrien Du Perron de Castera, *The Lady's Philosopher's Stone; Or, The Caprices of Love and Destiny: An Historical Novel. Written in French by M. L'Abbé de Castera; and now translated into English* (London: D. Browne, 1725). Leah Orr contends that this text cannot be attributed to Haywood on anything except a speculative basis, but most scholars accept the attribution for now. Leah Orr, "The Basis for Attribution in the Canon of Eliza Haywood," *The Library: The Transactions of the Bibliographical Society* 12, no. 4 (December 2011): 346–47, http://muse.jhu.edu/journals/lbt/summary/v012/12.4.orr.html.

12. Eliza Haywood, *Love In Excess*, ed. David Oakleaf, 2nd ed. (Toronto: Broadview, 2000), 100.

13. Eliza Haywood, *The Tea-Table: Or, A Conversation between some Polite Persons of both Sexes...Part the Second*, in *Selected Works of Eliza Haywood*, ed. Alexander Pettit, vol. 1, *Miscellaneous Writings, 1725–43* (London: Pickering & Chatto, 2000), 40, 41. All further references to *Part 2* are to this text. For more discussion of Behn's "To Daphnis," please see chapter 2.

14. Juliette Merritt, "Reforming the Coquet? Eliza Haywood's Vision of a Female Epistemology," in Wright and Newman, *Fair Philosopher*, 190.

15. See, for example, Drury, "Haywood's Thinking Machines"; Rebecca P. Bocchicchio, "'Blushing, Trembling, and Incapable of Defense': The Hysterics of *The British Recluse*," in Saxton and Bocchicchio, *The Passionate Fictions of Eliza Haywood*, 95–114; Rachel K. Carnell, "The Very Scandal of Her Tea Table: Eliza Haywood's Response to the Whig Sphere," in *Presenting Gender: Changing Sex in Early-Modern Culture*, ed. Chris Mounsey (Lewisburg: Bucknell University Press, 2001), 255–73. Potter's analysis of Haywood's handling of libertines in

her narratives implies a similar concern with the specifically male self, as well. Potter, "Eliza Haywood's Libertine Consistency."

16. For more discussion of this debate within philosophy, please see chapter 1.

17. Gerrard, *Aaron Hill*, 76–77; Kathryn R. King, *A Political Biography of Eliza Haywood* (London: Pickering & Chatto, 2012), 31–32; Patrick Spedding, *A Bibliography of Eliza Haywood* (London: Pickering & Chatto, 2004), 226.

18. King, *Eliza Haywood*, 31–32; Spedding, *Bibliography*, 226.

19. Eliza Haywood, *The Tea-Table: Or, A Conversation Between Some Polite Persons of Both Sexes* in *"Fantomina" and Other Works,* ed. Alexander Pettit, Margaret Case Croskery, and Anna C. Patchias (Toronto: Broadview Press, 2004), 73. All further references to *The Tea-Table* are to this text.

20. Like Dryden, John Donne was concerned with the larger significance of Galileo's revelations about nature. In "An Anatomy of the World: The First Anniversary," Donne writes, "And new philosophy calls all in doubt, / The element of fire is quite put out; / The sun is lost, and th'earth, and no man's wit / Can well direct him where to look for it. / And freely men confess that this world's spent, / When in the planets, and the firmament / They seek so many new; / They see that this / Is crumbled out again to his atomies." John Donne, "An Anatomy of the World: The First Anniversary," in *John Donne: The Complete Poems,* ed. A. J. Smith (New York: Penguin, 1971), l. 205–12.

21. Peter Dear, *Discipline and Experience: The Mathematical Way in the Scientific Revolution* (Chicago: University of Chicago Press, 1995), 6.

22. Samuel Johnson's *Dictionary of the English Language* shares a definition with the OED: "to range at large, enlarge upon." Samuel Johnson, *A Dictionary of the English Language....* (London: J. Knapton, C. Hitch and L. Hawes, A. Millar, R. and J. Dodsley, and M. and T. Longman, 1766), 1:np.

23. Until recently, scholars have regarded Haywood as an opponent of Walpole and the Whigs, but Kathryn R. King's *Political Biography of Eliza Haywood* proposes a more complex view of Haywood's political sympathies. The word "cabal," whatever the political affiliation, however, was certainly a politically loaded one in the 1720s. Rachel Carnell notes that it was associated both with Jacobitism and "women's tea tables," and that later Haywood would use "cabal" disparagingly and compare it unfavorably to a "league." Carnell, "Scandal of Her Tea Table," 256; Rachel Carnell, "It's Not Easy Being Green: Gender and Friendship in Eliza Haywood's Political Periodicals," in "Politics of Friendship," special issue, *Eighteenth-Century Studies* 32, no. 2 (Winter 1998/1999): 199–214, http://www.jstor.org/stable/30054219. For examples of the more traditional view of Haywood's opposition to Walpole, see Kirsten T. Saxton, introduction to Saxton and Bocchicchio, *The Passionate Fictions of Eliza Haywood*, 3; Lynn Marie Wright and Donald J. Newman, introduction to Wright and Newman, *Fair*

Philosopher, 28–29; Bannet, "Haywood's Spectator," 82–103; Kathryn R. King, "Patriot or Opportunist? Eliza Haywood and the Politics of *The Female Spectator*," in Wright and Newman, *Fair Philosopher*, 104–21; Earla A. Wilputte, "'Too ticklish to meddle with': The Silencing of *The Female Spectator*'s Political Correspondents," in Wright and Newman, *Fair Philosopher*, 122–40.

24. Ruth Gilbert, "Seeing and Knowing: Science, Pornography and Early Modern Hermaphrodites," in *At the Borders of the Human: Beasts, Bodies and Natural Philosophy in the Early Modern Period*, ed. Erica Fudge, Ruth Gilbert, and Susan Wiseman (New York: St. Martin's Press, 1999), 150–70.

25. Alexander Pettit, "Adventures in Pornographic Places: Eliza Haywood's *Tea-Table* and the Decentering of Moral Argument," *Papers on Language & Literature* 38, no. 3 (Summer 2002): 245, 247.

26. Bannet, "Haywood's Spectator," 82–103.

27. Ibid., 93–96.

28. Ibid., 93.

29. Ibid., 93–94.

30. Merritt, *Beyond Spectacle*, 11.

31. Aaron Hill, *The Plain Dealer* 11 (27 April 1724), in *The Plain Dealer: Being Select Essays on Several Curious Subjects....* vol. 1 (London: S. Richardson and A. Wilde, 1730): 78–82.

32. Gerrard, *Aaron Hill*, 76–77.

33. Pettit, "Adventures in Pornographic Places," 245; Alexander Pettit, Margaret Case Croskery, and Anna C. Patchias, Introduction to Pettit, Croskey, and Patchias,"*Fantomina*" and *Other Works*, 26.

34. Steven Shapin and Simon Shaffer, *Leviathan and the Air Pump: Hobbes, Boyle, and the Experimental Life. Including and Translation of Thomas Hobbes, "Dialogus Physicus de Natura Aeris," by Simon Shaffer* (Princeton: Princeton University Press, 1985), 20, 107.

35. Carnell, "Gender and Friendship," 204–5.

36. Merritt, *Beyond Spectacle*, 10–11.

37. A number of critics have pointed to ways in which Haywood was interested in Hobbes' ideas, among them Joseph Drury, "Haywood's Thinking Machines," and Helen Thompson, "Plotting Materialism: W. Charleton's *The Ephesian Matron*, E. Haywood's *Fantomina*, and Feminine Consistency," *Eighteenth-Century Studies* 35, no. 2 (Winter 2002): 195–214, doi:10.1353/ecs.2002.0017.

38. Lubey, "Eliza Haywood's Amatory Aesthetic," 310, original italics.

39. Eliza Haywood, *The Female Spectator*, in *Selections from "The Female Spectator*," ed. Patricia Meyers Spacks (New York: Oxford University Press, 1999), I, IV, 60.

40. Joseph Addison, *The Spectator* no. 1 (1 March 1711), in *Selections from "The Tatler" and "The Spectator*," ed. Angus Ross (New York: Penguin, 1982), 200.

41. Backscheider, "The Story," 29–41; Bannet, "Female Spectator"; Lubey, "Eliza Haywood's Amatory Aesthetic"; Hultquist, "Marriage in Haywood"; Cajka, "Unprotected Woman."
42. Lubey, "Eliza Haywood's Amatory Aesthetic."
43. Dear, *Discipline and Experience*, 6.
44. King, *Eliza Haywood*, 31–32.
45. Kathryn R. King, "Spying Upon the Conjurer: Haywood, Curiosity, and 'the Novel' in the 1720s," *Studies in the Novel* 30, no. 2 (Summer 1998): 184.
46. King, "Spying," 185–88.
47. Orr, "The Canon of Eliza Haywood," 350.
48. Haywood, *Part 2*, 44.
49. Pettit, Croskery and Patchias, Introduction, 26.
50. Pettit offers an opposite assessment of the erotics of *The Tea-Table* in "Adventures in Pornographic Places."
51. Most critics who read *The Tea-Table* treat it as interchangeable with *The Tea-Table Part 2* or as a seamless pair. Pettit argues that Haywood's agenda and use of space in both parts is contiguous; his analysis shifts between the two parts almost without acknowledging which passage comes from which text (Pettit, "Adventures in Pornographic Places"). King talks about "the *Tea Table* pamphlets" as one unit pursuing one agenda (King, *Eliza Haywood*, 32). Spedding's discussion of *The Tea-Table* refers to the poem about Hillarius although this poem only appears in *Part 2* (Spedding, *Bibliography*, 226).
52. Steven Shapin, *Never Pure: Historical Studies of Science as If It Was produced by People with Bodies, Situated in Time, Space, Culture, and Society, and Struggling for Credibility and Authority* (Baltimore: The Johns Hopkins University Press, 2010), 238.
53. See also Pettit, "Adventures in Pornographic Places," 255.
54. Drury, "Haywood's Thinking Machines," 215–17. In this point my argument directly opposes Sharon Harrow's claim that Haywood's poetry and novels draw on "Enlightenment individualism" to justify the independent acting of amatory heroines. Sharon Harrow, "Having Text: Desire and Language in Haywood's *Love in Excess* and *The Distressed Orphan*," *Eighteenth-Century Fiction* 22, no. 2 (Winter 2009–10): 279–308.
55. See, for example, Susan Paterson Glover, *Engendering Legitimacy: Law, Property, and Early Eighteenth-Century Fiction* (Lewisburg: Bucknell University Press, 2006).

5   The Moral Observer

1. Susan Staves, *A Literary History of Women's Writing in Britain, 1660–1789* (New York: Cambridge University Press, 2008), 186; William H.

McBurney, "Mrs. Mary Davys: Forerunner of Fielding," *PMLA* 74, no. 4 (September 1959): 348–55, http://www.jstor.org/stable/460444; Jean B. Kern, "Mary Davys as Novelist of Manners." *Essays in Literature* 10, no. 1 (Spring 1983): 29–38; Victoria Joule, "Mary Davys's Novel Contribution to Women and Realism," *Women's Writing* 17, no. 1 (May 2010): 30–48, doi:10.1080/09699080903533262; Jane Spencer, *The Rise of the Woman Novelist: From Aphra Behn to Jane Austen* (Oxford: Blackwell, 1986), 143–47; Janet Todd, *The Sign of Angellica: Women, Writing and Fiction, 1660–1800* (New York: Columbia University Press, 1989), 50–51; Frans De Bruyn, "Mary Davys (1674–1732)," in *Dictionary of Literary Biography: British Novelists, 1660–1800*, ed. Philip Breed Dematteis and Leemon B. McHenry, vol. 39 (Detroit: Gale Research, 1985), 135–37; Martha Bowden, Introduction to *"The Reform'd Coquet," "Familiar Letters Betwixt a Gentleman and a Lady," and "The Accomplish'd Rake,"* ed. Martha F. Bowden (Lexington, KY: The University Press of Kentucky, 1999), xxv; Donald Hal Stefanson, "The Works of Mary Davys: A Critical Edition (Vol. I and II)" (PhD dissertation, University of Iowa, 1971), xiv–xv.

2. Like Victoria Joule, for example, Jesse Molesworth does not see women writers as part of the realist novel. Joule, "Mary Davys," 31; Jesse Molesworth, *Chance and the Eighteenth-Century Novel: Realism, Probability, Magic* (New York: Cambridge University Press, 2010). See also Staves, *Literary History*, 184.

3. Rachel Carnell, *Partisan Politics, Narrative Realism, and the Rise of the British Novel* (New York: Palgrave Macmillan, 2006). See also J. Paul Hunter, *Before Novels: The Cultural Contexts of Eighteenth-Century English Fiction* (New York: W. W. Norton, 1990), 56.

4. John Locke, *An Essay Concerning Human Understanding*, 6th ed. (Kitchener, ON: Batoche Books, 2001), IV.iv.7, eBook.

5. De Bruyn, "Mary Davys," 135. See also Todd, *Angellica*, 51.

6. Sarah Prescott, *Women, Authorship and Literary Culture, 1690–1740* (New York: Palgrave Macmillan, 2003), 44.

7. McBurney, "Mrs. Mary Davys," 355. See also Prescott, *Women, Authorship and Literary Culture*, 44; De Bruyn, "Mary Davys," 134–35; J. A. Downie, "Mary Davys's 'Probable Feign'd Stories' and Critical Shibboleths about 'The Rise of the Novel,'" *Eighteenth-Century Fiction* 12, no. 2–3 (January–April 2000): 309–26, doi:10.1353/ecf.2000.0033.

8. Mary Davys, Preface to *The Works of Mrs. Davys*, vol. 1 (London: H. Woodfall, 1725), 1:v.

9. Mary Davys, Dedication to *The Fugitive* (London: W. Sawbridge, 1705), np.

10. "A List of the Subscribers Names" to *The Reform'd Coquet* in Bowden, *"The Reform'd Coquet,"* 7; "A List of the Subscribers" to *The Works of Mrs. Davys* in Bowden, *"The Reform'd Coquet,"* 89.

11. Martha Bowden, "Silences, Contradictions, and the Urge to Fiction: Reflections on Writing about Mary Davys," *Studies in the Literary*

Imagination 36, no. 2 (2003): 127–47; Martha Bowden, "Mary Davys: Self-Presentation and Woman Writer's Reputation in the Early Eighteenth Century," *Women's Writing* 3, no. 1 (1996): 22–25; Bowden, Introduction, xvi–xvii; Peggy Keeran and Jennifer Bowers, *Literary Research and the British Eighteenth Century: Strategies and Sources* (Lanham, MD: Scarecrow Press, 2013), 256–62.

12. Jane Spencer, "Amatory and Scandal Fiction," in *The Oxford History of the English Novel*, ed. Thomas Keymer, vol. 1 (Oxford: Oxford University Press, forthcoming); Bowden, Introduction, xvii, xx; Martha Bowden, "Chronology of Events in the Life of Mary Davys," in Bowden, *"The Reform'd Coquet,"* xlvii–xlviii.

13. Recently, Gerd Bayer has proposed another possible narrative by Mary Davys: "A Gift and No Gift," published in *The Gentleman's Journal* in 1693. Although Bayer's proposal is persuasive, he acknowledges that the evidence is not definitive; consequently, "A Gift and No Gift" remains outside the scope of this study. Gerd Bayer, "A Possible Early Publication by Mary Davys and Its Swiftian Afterglow," *Notes & Queries* 59, no. 2 (Spring 2012): 194–97, doi:10.1093/notesj/gjs006.

14. Toni Bowers, *Force or Fraud: British Seduction Stories and the Problem of Resistance, 1660–1760* (Oxford: Oxford University Press, 2011); Alice Wakely, "Mary Davys and the Politics of Epistolary Form," in *"Cultures of Whiggism": New Essays on English Literature and Culture in the Long Eighteenth Century*, ed. David Womersley, assisted by Paddy Bullard and Abigail Williams (Newark, DE: University of Delaware Press, 2005), 257–67; Spencer, "Amatory and Scandal Fiction," np; Susan Glover, *Engendering Legitimacy: Law, Property, and Early Eighteenth-Century Fiction* (Lewisburg: Bucknell University Press, 2006), 82–85; Stefanson, "Mary Davys," xi–xii.

15. Spencer, "Amatory and Scandal Fiction," np.

16. Glover, *Engendering Legitimacy*, 85, 97.

17. Mary Davys, *Familiar Letters Betwixt a Gentleman and a Lady*, in Bowden, *"The Reform'd Coquet,"* 110. All further references to this text are abbreviated *FL*.

18. According to Richard Westfall, Flamsteed got the copies on March 28, 1714 and burned them after separating the materials he wanted from the dreck, but Westfall does not give an exact date for what Davys calls the "Conflagration." Presumably it took Flamsteed several days to dissect 300 copies, so Davys's dating the fire to April makes sense. Richard S. Westfall, *Never at Rest: A Biography of Isaac Newton* (Cambridge: Cambridge University Press, 1980), 694–96. See also Bowden, *"The Reform'd Coquet,"* 241n47.

19. Gary W. Kronk, *Cometology: A Catalog of Comets*, vol. 1: *Ancient-1799* (Cambridge: Cambridge University Press, 1999), 1:389–91.

20. William Whiston and Humphrey Dutton, *A New Method for Discovering the Longitude both at Sea and Land, Humbly Proposed to the Consideration of the Publick....* (London: John Phillips, 1714).

21. Westfall, *Never at Rest*, 835–36; Dava Sobel, *Longitude: The True Story of a Lone Genius Who Solved the Greatest Scientific Problem of His Time* (New York: Walker and Company, 1995), 53.

22. I am indebted to W. L. Gold, Lisa Rose Wiles, and most especially Mary Ann O'Donnell for their assistance in tracking down the source of Davys's references.

23. Michael Hunter, "Science and Astrology in Seventeenth-Century England: An Unpublished Polemic by John Flamsteed," in *Science and the Shape of Orthodoxy: Intellectual Change in Late Seventeenth-Century Britain* (Woodbridge: Boydell Press, 1995), 245–85; A. J. Meadows, "John Flamsteed, Our Astronomical Observator," *Notes and Records of the Royal Society of London*, 50, no. 2 (July 1996): 252, http://www.jstor.org/stable/531915; Jan Golinski, *British Weather and the Climate of Enlightenment* (Chicago: University of Chicago Press, 2011), 50–51.

24. Hunter, "Science and Astrology," 247–48.

25. George S. Rousseau, *Enlightenment Borders: Pre-And Post-Modern Discourses: Medical, Scientific* (Manchester, UK: Manchester University Press, 1991), 331.

26. *The Black-Day, or, a Prospect of Doomsday. Exemplified in the Great and Terrible Eclipse, Which Will Happen on Friday the 22d of April, 1715....*(London: J. Reid and R. Burleigh, 1715). See also, for example, J. Parker, *The History of Eclipses....*(London: J. Read and R. Burleigh, 1715).

27. *Oxford English Dictionary online*, s.v. "Witch, n.1," accessed March 4, 2013, http://www.oed.com/view/Entry/229574?p=emailAqMCyretGuG9k&d=229574. Mary Davys, *The Accomplish'd Rake* in Bowden, "*The Reform'd Coquet*," 167; Mary Davys, *The Lady's Tale* in *The Works of Mrs. Davys*, vol. 2 (London: H. Woodfall, 1725), 134. All further references to *The Accomplish'd Rake* are abbreviated *AR*; all further references to *The Lady's Tale* are abbreviated *LT*.

28. Victoria Joule, for example, suggests that Sir John Galliard is an "homage to the 'Johnians,'" the men of St. John's College at Cambridge, for their "helpful exchange of ideas" and suggestions for the manuscript of *The Accomplish'd Rake*. Joule, "Mary Davys," 37. See also Lindy Riley, "Mary Davys's Satiric Novel *Familiar Letters*: Refusing Patriarchal Inscription of Women," in *Cutting Edges: Postmodern Critical Essays on Eighteenth-Century Satire*, ed. James E. Gill (Knoxville: University of Tennessee Press, 1995), 207; Glover, *Engendering Property*, 85–86. Davys herself admitted that Cambridge students encouraged her to write but rejected the idea that they had a hand in the composition. Mary Davys, Preface to *The Works of Mrs. Davys*, vol. 2 (London: H. Woodfall, 1725), 2:7–8.

29. Rousseau, *Enlightenment Borders*, 275–77. Marjorie Hope Nicolson notes that women in mid-century Dublin could attend lectures on Newtonian optics as well, although it is not clear whether those lectures

would have been held in coffeehouses. Marjorie Hope Nicolson, *Newton Demands the Muse: Newton's "Opticks" and the Eighteenth-Century Poets* (Princeton: Princeton University Press, 1946), 16.

30. Spencer, "Amatory and Scandal Fiction," np; Wakely, "Mary Davys," 264–65.

31. Eliza Haywood translated a similar event in *The Lady's Philosopher's Stone* (1725), a text also interested in alchemy and irrationality. Louis-Adrien Du Perron de Castera, *The Lady's Philosopher's Stone; Or, the Caprices of Love and Destiny: An Historical Novel*. Written in French by M. L'Abbé de Castera; and now translated into English (London: D. Browne and S. Chapman, 1725), 11–13.

32. Spencer also observes the political dimension of Artander's observation. Spencer, "Amatory and Scandal Fiction," np.

33. For more discussion of tea tables, please see chapter 4.

34. Carol Pateman, *The Sexual Contract* (Stanford: Stanford University Press, 1988).

35. For more discussion of Jane Barker and the representation of incomplete knowing through framed narrative, please see chapter 3.

36. Riley, "Mary Davys's Satiric Novel," 218–19.

37. See also Riley, "Mary Davys's Satiric Novel."

38. Although Bowden at one point agreed with McBurney that *The False Friend* was a pirated version of *The Cousins*, she has since come to agree with De Bruyn and Stefanson that it was published with Davys's permission. Bowden, "Silences," 137; McBurney, "Mrs. Mary Davys," 354n; De Bruyn, "Mary Davys," 13; Stefanson, "Mary Davys," xxxii–xxxiv.

39. Mary Davys, *The Cousins; A Novel* in *The Works of Mrs Davys*, vol. 2 (London: H. Woodfall, 1725), 2:211–14, 2:215–17.

40. Daniel P. Gunn, "Free Indirect Discourse and Narrative Authority in *Emma*," *Narrative* 12, no. 1 (January 2004): 35.

41. Laura Buchholtz, "The Morphing Metaphor and the Question of Narrative Voice," *Narrative* 17, no. 2 (May 2009): 200–219.

42. Stefanson points out that the primary difference between originals and those published in *Works* resides in "accidentals" such as spelling and punctuation (Stefanson, "Mary Davys," xxxii–xxxiv).

43. Bowden, "Silences," 127–47; Bowden, "Mary Davys," 22–25; Bowden, Introduction, xvi–xvii. For a discussion of *The Fugitive* as autobiographical, see, for example, Joule, "Mary Davys," 30–48; De Bruyn, "Mary Davys," 132; Stefanson, "Critical Edition," xxi. Jean Kern and William McBurney read all of Davys's novels as literal or emotional autobiographies. Kern, "Mary Davys," 29–38; McBurney, "Mrs. Mary Davys," 348–55.

44. Mary Davys, *The Amours of Alcippus and Lucippe. A Novel* (London: James Round, 1704), 1. All further references to this text are abbreviated *A&L*.

45. Mary Davys, Dedication to *The Amours of Alcippus and Lucippe: A Novel Written by a Lady* (London: James Round, 1704), np.

46. Bowden, Introduction, xiv.
47. Mary Davys, Preface to *The Amours of Alcippus and Lucippe: A Novel Written by a Lady* (London: James Round, 1704), np.
48. Davys claims to have begun *Alcippus and Lucippe* in 1700, which may suggest that she wrote it in Ireland where Margaret Walker, as the daughter of John Jeffreyson, appointed to a judgeship in Dublin by William and Mary, would possibly have been more of a political and social presence. That Davys retained the dedication for a 1704 publication in England is interesting. While there is no knowing for certain why she did so, the fact that Margaret Walker was not a powerful person in the English sociopolitical scene suggests that even in dedicating the novel to a person without a presence, so to speak, Davys was refusing to frame or contextualize the empirical aspects of the narrative. Bowden, Introduction, xiii.
49. Davys expressed this Anglocentric view in other texts, as well. In both *The Fugitive* and in *The Merry Wanderer*, the narrator proclaims proudly that although "I have never been at France for new Fashions, nor at Rome for Religion of a Song, yet I hope England is not so barren of diversion, but one may pick up some things in it, worthy of Note." This passage is the same in *The Merry Wanderer* although other parts of the narrative were changed. In the Dedication to the second volume of the *Works*, Davys appeals again to nationalist fervor: "I believe every body will join with my Opinion, that English Ladies are the most accomplish'd Women in the World." Mary Davys, *The Fugitive* (London: G. Sawbridge, 1705), 1; Mary Davys, *The Merry Wanderer* in *The Works of Mrs. Davys*, vol. 1 (London: H. Woodfall, 1725), 1:161; Mary Davys, Dedication to *The Works of Mrs. Davys*, vol. 2 (London: H. Woodfall, 1725), 2:4. All further references to *The Merry Wanderer* are abbreviated *MW* in the text.
50. Davys revised "Alcippus" in *Alcippus and Lucippe* to "Alcipus" in *The Lady's Tale*.
51. Davys, Preface to *Works*, 1:vi.
52. For more discussion of widows in the eighteenth-century novel, see Karen Bloom Gevirtz, *Life After Death: Widows and the English Novel, Defoe to Austen* (Newark, DE: University of Delaware Press, 2005).
53. Because the texts are structurally the same and the prose extremely similar, I am quoting from *The Merry Wanderer* unless a difference between the texts requires a passage from *The Fugitive*. The narrator's opening autobiographical declarations are significantly different, for example, even though the technique of self-introduction and the first-person point of view are the same. Although *The Fugitive* is valuable because it is an early expression of Davys's interests, I am using *The Merry Wanderer* because it was not revised when it could have been, indicating Davys's commitment to the text. Davys, *The Merry Wanderer*, 1:163–69.

54. McBurney's pronouncement on this topic has been generally accepted. In addition to McBurney, "Mrs. Mary Davys," 350–53, see also, for example, Staves, *Literary History*, 184; De Bruyn, "Mary Davys ," 134–35; Glover, *Engendering Legitimacy* 82; Bowden, introduction, xxvii-xxxi; Kern, "Mary Davys," 33–35; Stefanson, "Mary Davys," xviii.
55. Bowden, introduction, xviii–xix; Bowden, "Chronology," xlvii.
56. Mary Davys, *The Reform'd Coquet*, in Bowden, "*The Reform'd Coquet*," 11. All further references to this text are abbreviated *RC*.
57. De Bruyn, "Mary Davys," 137.
58. A search on ECCO for "compound" almost exclusively yields books on mathematics until about 1700, when medical books using the term suddenly begin to appear. The Royal College of Physicians, for example, published William Salmon's *Pharmacopoia Londinensis. Or, the new London dispensatory* in 1702, which explained the method of compounding medicines and provided an extensive "pharmacopoia" of recipes for the "Publick Good" and "fitted to the meanest Capacity." In the same year, Robert Pitt published *The craft and frauds of physick expos'd* which promised to reveal "The costly preparations now in greatest esteem, condemn'd.... With Instructions to Prevent being Cheated and Destroy'd by the prevailing Practice." John Pechey's *Compleat Herbal of Physical Plants* (1707) provided recipes for making medicinal compounds from plants rather than chemicals, and George Smith's *A compleat body of distilling* (1722), designed for household use, explained the chemistry of distilling. "Abstract" and "quantum" also had their usages from natural philosophy. Primarily associated with philosophy in the seventeenth and early eighteenth centuries, "quantum" referred to something with a measurable quantity, while "abstract," a process by which one substance was removed from another, became associated with chemistry in the mid-seventeenth century. *Oxford English Dictionary Online*, s.v. "compound, n.1," accessed April 29, 2013; http://www.oed.com.ezproxy.shu.edu/view/Entry/37831?rskey =56NZG8&result=1&isAdvanced=false; *Oxford English Dictionary Online*, s.v. "quantum, n. and adj.," accessed April 29, 2013, http:// www.oed.com. ezproxy.shu.edu/view/Entry/155941?redirectedFrom =quantum; *Oxford English Dictionary Online*, s.v. "abstract, v.," accessed April 29, 2013 http://www.oed.com.ezproxy.shu.edu/view/ Entry/759; William Salmon, *Pharmacopoia Londinensis. Or, the New London Dispensatory....* (London: Royal College of Physicians, 1702); Robert Pitt, *The Craft and Frauds of Physick Expos'd....* (London: Tim Childe, 1702); John Pechey, *The Compleat Herbal of Physical Plants....* (London: R. and J. Bonwicke, 1707); George Smith, *A Compleat Body of Distilling, Explaining the Mysteries of that Science, in a Most Easy and Familiar Manner....* (London: Bernard Lintot, 1725).
59. De Bruyn, "Mary Davys," 135.

60. Glover, *Engendering Legitimacy*, 87.
61. Ibid., 134.
62. Davys uses the idea of movement and freedom at the end of several of her novels. The Irish widow is still peripatetic when *The Fugitive* and *The Merry Widow* conclude, and the untamed Abaliza walks out of the house in *The Lady's Tale*. The couples end *The Cousins* and *Alcippus and Lucippe* by coming to rest, however, suggesting that Davys associates the conventional ending-by-marriage with stasis.
63. Spencer, *Rise*, 145–47.
64. Staves, *Literary History*, 186.
65. Critics generally credit the stage with sharpening Davys's satiric skills, but considering that *The Fugitive* and *Alcippus and Lucippe* were both written well before she had a play staged (*The Northern Heiress*, 1716), there is likely a different or an additional explanation. Not all Augustan satirists were playwrights, after all, aspiring or otherwise. See, for example, Riley, "Mary Davys," 207; Wakely, "Mary Davys," 267; De Bruyn, "Mary Davys," 134–35.
66. Lorraine Daston and Peter Galison, *Objectivity*, paperback ed. (New York: Zone Books, 2010).

Conclusion

1. Jonah Lehrer, *Proust Was a Neuroscientist* (New York: Houghton Mifflin, 2004), xii.
2. Paul Dawson, "The Return of Omniscience in Contemporary Fiction," *Narrative* 17, no. 2 (May 2009): 143–61, doi:10.1353/nar.0.0023; Susan Sniader Lanser, *Fictions of Authority: Women Writers and Narrative Voice* (Ithaca, NY: Cornell University Press, 1992); Susan Sniader Lanser, *The Narrative Act: Point of View in Prose Fiction* (Princeton: Princeton University Press, 1981).
3. Dawson, "The Return of Omniscience," 149.
4. Ibid., 144.
5. J. Paul Hunter, *Before Novels: The Cultural Contexts of Eighteenth-Century English Fiction* (New York: W. W. Norton, 1990), 45.
6. See, for example, Jonathan Culler, "Omniscience," *Narrative* 12, no. 1 (January 2004): 22–34.
7. Svetlana Alpers, *The Art of Describing: Dutch Art in the Seventeenth Century* (Chicago: University of Chicago Press, 1983), 5.
8. Marjorie Hope Nicolson, *Newton Demands the Muse: Newton's "Opticks" and the Eighteenth-Century Poets* (Princeton: Princeton University Press, 1946); Mordechai Feingold, *The Newtonian Moment: Isaac Newton and the Making of Modern Culture* (New York: The New York Public Library and Oxford University Press, 2004); Laura Baudot, "An Air of History: Joseph Wright's and Robert Boyle's Air

Pump Narratives," *Eighteenth-Century Studies* 46, no. 1 (Fall 2012): 1–28; Lorraine Daston and Peter Galison, *Objectivity*, paperback ed. (New York: Zone Books, 2010).

9. For more comparison of Austen's technique in her early drafts and her later publications, see, for example, Narelle Shaw, "Free Indirect Speech and Jane Austen's 1816 Revision of *Northanger Abbey*," *Studies in English Literature, 1500–1900* 30, no. 4 (Autumn 1990): 591, http://www.jstor.org/stable/450561.

Bibliography

Addison, Joseph. *The Spectator no. 1* (1 March 1711). In *Selections from "The Tatler" and "The Spectator"*, edited by Angus Ross, 197–200. New York: Penguin, 1982.

Ahern, Stephen. "'Glorious ruine': Romantic Excess and the Politics of Sensibility in Aphra Behn's *Love-Letters*." *Restoration* 29, no. 1 (Spring 2005): 29–45.

Alcoff, Linda and Elizabeth Potter, eds. *Feminist Epistemologies*. New York: Routledge, 1993.

Allestree, Richard. *The Ladies Calling*. Oxford, 1673.

Alpers, Svetlana. *The Art of Describing: Dutch Art in the Seventeenth Century*. Chicago: University of Chicago Press, 1983.

Altegoer, Diana B. *Reckoning Words: Baconian Science and the Construction of Truth in English Renaissance Culture*. Madison: Fairleigh Dickinson University Press, 2000.

Anderson, Emily Hodgson. "Novelty in Novels: A Look at What's New in Aphra Behn's *Oroonoko*." *Studies in the Novel* 39, no. 1 (Spring 2007): 1–16.

Anderson, Misty G. "Tactile Places: Materializing Desire in Margaret Cavendish and Jane Barker." *Textual Practice* 13, no. 2 (1999): 329–52.

Armstrong, Nancy. *Desire and Domestic Fiction: A Political History of the Novel*. New York: Oxford University Press, 1987.

Atkinson, Dwight. *Scientific Discourse in Sociohistorical Context: "The Philosophical Transactions of the Royal Society of London," 1675–1975*. Mahwah, NJ: Lawrence Erlbaum Associates, 1999.

Backscheider, Paula, ed. *Revising Women: Eighteenth-Century Women's Fiction and Social Engagement*. Baltimore: The Johns Hopkins University Press, 2000.

———. "The Shadow of an Author: Eliza Haywood." *Eighteenth-Century Fiction* 11, no. 1 (October 1998): 79–102. http://digitalcommons.mcmaster .ca/ecf/vol11/iss1/1.

———. "The Story of Eliza Haywood's Novels." In Saxton and Bocchicchio, *The Passionate Fictions of Eliza Haywood*, 19–47.

Backscheider, Paula R. and John J. Richetti, eds. *Popular Fiction by Women 1660–1730: An Anthology*. Oxford: Oxford University Press, 1996.

Bacon, Francis. *The Great Instauration*. In Sargent, *Francis Bacon*, 66–89.

———. *The New Atlantis*. In Sargent, *Francis Bacon*, 239–68.

———. *The New Organon*. In *The Complete Essays of Francis Bacon*, edited by Henry LeRoy Finch, 179–266. New York: Washington Square Press, 1963.

Bacon, Francis. *Preparative to the New Organon.* In Sargent, *Francis Bacon,* 190–206.

Ballaster, Ros. *Seductive Forms: Women's Amatory Fiction from 1684 to 1740.* Oxford: Clarendon Press, 1998.

———. "Taking Liberties: Revisiting Behn's Libertinism." *Women's Writing* 19, no. 2 (May 2012): 165–76. http://dx.doi.org/10.1080/09699082.201 1.646861.

Bannet, Eve Tavor. "Haywood's Spectator and the Female World." In Wright and Newman, *Fair Philosopher,* 82–103.

Barker, Jane. "A Farewell to Poetry, with a long Digression on Anatomy." In *Poetical Recreations: Consisting of Original Poems, Songs, Odes etc; With several New Translations. In Two Parts: Part I. Occasionally written by Mrs. Jane Barker. Part II. By Several Gentlemen of the Universities, and Others.* London: Benjamin Crayle, 1688.

———. "An Invitation to my Friends at Cambridge." In *Poetical Recreations.*

———. *A Lining for the Patch-Work Screen for the Ladies.* In Wilson, *The Galesia Trilogy,* 175–290.

———. "Love Intrigues." In Wilson, *The Galesia Trilogy,* 1–47.

———. "On my Mother and my Lady W – – who both lay sick at the same time Under the Hands of Dr. Paman." In *Poetical Recreations.*

———. *A Patch-Work Screen for the Ladies.* In Wilson, *The Galesia Trilogy,* 49–173.

Barnes, Diana. "Familiar Epistolary Philosophy: Margaret Cavendish's *Philosophical Letters* (1664)." *Parergon* 26, no. 2 (2009): 39–64.

———. "The Restoration of Royalist Form in Margaret Cavendish's *Sociable Letters.*" In *Women Writing 1550–1750,* edited by Jo Wallwork and Paul Salzman, special issue of *Meridian: The La Trobe University English Review* 18, no. 1 (2001): 201–14.

Battigelli, Anna. *Margaret Cavendish and the Exiles of the Mind.* Lexington, KY: University Press of Kentucky, 1998.

Baudot, Laura. "An Air of History: Joseph Wright's and Robert Boyle's Air Pump Narratives." *Eighteenth-Century Studies* 46, no. 1 (Fall 2012): 1–28.

Bayer, Gerd. "A Possible Early Publication by Mary Davys and Its Swiftian Afterglow." *Notes & Queries* 59, no. 2 (Spring 2012): 194–97.

Beebee, Thomas O. *Epistolary Fiction in Europe: 1500–1850.* Cambridge: Cambridge University Press, 1999.

Behn, Aphra. "The Adventure of the Black Lady." In *The Works of Aphra Behn,* edited by Janet Todd, 315–20. Vol. 3: *"The Fair Jilt" and Other Stories.* Columbus: Ohio State University Press, 1995.

———. "The History of the Nun." In Todd, *The Works of Aphra Behn,* 3:211–58.

———. *Love-Letters Between a Nobleman and His Sister.* In *The Works of Aphra Behn,* edited by Janet Todd, 3–439. Vol. 2: *Love-Letters Between a Nobleman and His Sister.* Columbus: Ohio State University Press, 1993.

———. *Oroonoko.* Edited by Joanna Lipking, 5–65. New York: W. W. Norton, 1997.

———. *The Roundheads.* In *The Works of Aphra Behn,* edited by Janet Todd, 357–424. Vol. 6: *The Plays, 1678–1682.* Columbus: Ohio State University Press, 1993.

———. "To the Unknown Daphnis on his Excellent Translation of Lucretius." In *The Works of Aphra Behn,* edited by Janet Todd, 25–28. Vol. 1: *Poetry.* Columbus: Ohio State University Press, 1992.

———. "The Unfortunate Happy Lady." In Todd, *The Works of Aphra Behn,* 3:361–87.

Bender, John. *Imagining the Penitentiary: Fiction and the Architecture of Mind in Eighteenth-Century England.* Chicago: University of Chicago Press, 1987.

Benedict, Barbara M. "The Curious Genre: Female Inquiry in Amatory Fiction." *Studies in the Novel* 30, no. 2 (Summer 1998): 194–210.

———. "The Mad Scientist: The Creation of a Literary Stereotype." In *Imagining the Sciences: Expressions of New Knowledge in the "Long" Eighteenth Century,* edited by Robert C. Leitz III and Kevin L. Cope, 68–70. New York: AMS Press, 2004.

———. "The Spirit of Things." In *The Secret Life of Things: Animals, Objects, and It-Narratives in Eighteenth-Century England,* edited by Mark Blackwell, 19–42. Lewisburg: Bucknell University Press, 2007.

Bensaude-Vincent, Bernadette and Christine Blondel, eds. *Science and Spectacle in the European Enlightenment.* Burlington: Ashgate, 2008.

The Black-Day, or, a Prospect of Doomsday. Exemplified in the Great and Terrible Eclipse, Which Will Happen on Friday the 22d of April, 1715.... London: J. Reid and R. Burleigh, 1715.

Blackwell, Mark, ed. *The Secret Life of Things: Animals, Objects, and It-Narratives in Eighteenth-Century England.* Lewisburg: Bucknell University Press, 2007.

Bocchicchio, Rebecca P. "'Blushing, Trembling, and Incapable of Defense': The Hysterics of *The British Recluse.*" In Saxton and Bocchicchio, *The Passionate Fictions of Eliza Haywood,* 95–114.

Boone, Troy. "Narrating the Apparition: Glanvill, Defoe, and the Rise of Gothic Fiction." *The Eighteenth Century* 35, no. 2 (1994): 173–89.

Bowden, Martha. "Chronology of Events in the Life of Mary Davys." In Bowden, *"The Reform'd Coquet,"* xlvii–xlviii.

———. Introduction to Bowden, *"The Reform'd Coquet,"* ix–xlvi.

———. "Mary Davys: Self-Presentation and Woman Writer's Reputation in the Early Eighteenth Century." *Women's Writing* 3, no. 1 (1996): 17–33.

———, ed. *"The Reform'd Coquet," "Familiar Letters Betwixt a Gentleman and a Lady,"* and *"The Accomplish'd Rake"* by Mary Davys. Lexington, KY: The University Press of Kentucky, 1999.

———. "Silences, Contradictions, and the Urge to Fiction: Reflections on Writing about Mary Davys." *Studies in the Literary Imagination* 36, no. 2 (2003): 127–47.

Bower, Anne L. "Dear —: In Search of New (Old) Forms of Critical Address." In Gilroy and Verhoeven, *Epistolary Histories,* 155–75.

Bowers, Toni. *Force or Fraud: British Seduction Stories and the Problem of Resistance, 1660–1760.* Oxford: Oxford University Press, 2011.

———. "Sex, Lies and Invisibility: Amatory Fiction from the Restoration to Mid-Century." In *The Columbia History of the British Novel,* edited by John Richetti, 50–72. New York: Columbia University Press, 1994.

Boyle, Robert. *The Christian Virtuoso I.* In *The Works of Robert Boyle,* edited by Michael Hunter and Edward B. Davis, 281–327. Vol. 11, *"The Christian Virtuoso" and Other Publications of 1687–91.* London: Pickering & Chatto, 2000.

———. *New Experiments and Observations Touching Cold, Or an Experimental History of Cold.*... In *The Works of Robert Boyle,* edited by Michael Hunter and Edward B. Davis, 203–517. Vol. 4, *"Colours" and "Cold," 1664–5.* London: Pickering & Chatto, 1999.

———. *Some Considerations Touching the Usefulness of Experimental Natural Philosophy. The First Part.* In *The Works of Robert Boyle,* edited by Michael Hunter and Edward B. Davis, 189–290. Vol. 3: *"The Usefulness of Natural Philosophy" and sequels to "The Spring of the Air," 1662–3.* London: Pickering & Chatto, 1999.

Bratach, Anne. "Following the Intrigue: Aphra Behn, Genre, and Restoration Science." *Journal of Narrative Technique* 26, no. 3 (Fall 1996): 209–27.

Broad, Jacqueline. "Mary Astell's Machiavellian Moment? Politics and Feminism in *Moderation truly Stated.*" In Wallwork and Salzman, *Early Modern Englishwomen Testing Ideas,* 9–23.

Brown, Laura. "Feminization of Ideology: Form and the Female in the Long Eighteenth Century." In *Ideology and Form in Eighteenth-Century Literature,* edited by David H. Richter, 223–40. Lubbock: Texas Tech University Press, 1999.

Buchholtz, Laura. "The Morphing Metaphor and the Question of Narrative Voice." *Narrative* 17, no. 2 (May 2009): 200–219.

Burnet, Thomas. *Remarks upon an Essay concerning Humane Understanding, in a Letter to the Author.* London: M. Wotton, 1697.

Butler, Joseph. *The Analogy of Religion, Natural and Revealed.*... Dublin: J. Jones, 1736.

Butler, Judith. *Bodies that Matter: On the Discursive Limits of 'Sex.'* New York: Routledge, 1993.

———. *Gender Trouble: Feminism and the Subversion of Identity.* New York: Routledge, 1990.

Cajka, Karen. "The Unprotected Woman in Eliza Haywood's *The History of Jemmy and Jenny Jessamy.*" In Carlile, *Masters,* 47–58.

Carlile, Susan, ed. *Masters of the Marketplace: British Women Novelists of the 1750s.* Bethlehem: Lehigh University Press, 2011.

Carnell, Rachel. "It's Not Easy Being Green: Gender and Friendship in Eliza Haywood's Political Periodicals." Special issue, "Politics of Friendship," *Eighteenth-Century Studies* 32, no. 2 (Winter 1998/1999): 199–214. http://www.jstor.org/stable/30054219.

————. *Partisan Politics, Narrative Realism, and the Rise of the British Novel*. New York: Palgrave Macmillan, 2006.

————. "The Very Scandal of Her Tea Table: Eliza Haywood's Response to the Whig Sphere." In *Presenting Gender: Changing Sex in Early-Modern Culture*, edited by Chris Mounsey, 255–73. Lewisburg: Bucknell University Press, 2001.

Carter, Philip. *Men and the Emergence of Polite Society, Britain, 1660–1800*. Harlow: Pearson, 2001.

de Castera, Louis-Adrien Du Perron. *The lady's philosopher's stone; or, the caprices of love and destiny: an historical novel. Written in French by M. L'Abbé de Castera; and now translated into English*. London: D. Browne and S. Chapman, 1725.

Chibka, Robert L. "'Oh! Do Not Fear a Woman's Invention': Truth, Falsehood, and Fiction in Aphra Behn's *Oroonoko*." *Texas Studies in Literature and Language* 30, no. 4 (Winter 1998): 510–37.

Chico, Tita. "Gimcrack's Legacy: Sex, Wealth, and the Theater of Experimental Philosophy." *Comparative Drama* 42, no. 2 (Spring 2008): 29–49.

Christianson, Gale E. *Isaac Newton*. New York: Oxford University Press, 2005. eBook.

Clark, Peter. *British Clubs and Societies, 1500–1800*. New York: Oxford University Press, 2000.

Clucas, Stephen. "Joanna Stephens's Medicine and the Experimental Philosophy." In Zinsser, *Men, Women, and the Birthing of Modern Science*, 141–58.

Cockburn, Catherine Trotter. *A letter to Dr. Holdsworth, occasioned by his sermon preached before the University of Oxford....* London: Benjamin Motte, 1726.

Code, Lorraine. "Taking Subjectivity into Account." In Alcoff and Potter, *Feminist Epistemologies*, 15–48.

Cohen, I. Bernard and Richard S. Westfall. General Introduction to *Newton: Texts, Backgrounds, Commentaries*, xi–xv. Edited by I. Bernard Cohen and Richard S. Westfall. New York: W. W. Norton, 1995.

Coppola, Al. "Retraining the Virtuoso's Gaze: Behn's *Emperor of the Moon*, The Royal Society, and the Spectacles of Science and Politics." *Eighteenth-Century Studies* 41, no. 4 (Summer 2008): 481–506.

Cotes, Roger. "Cotes's Preface to the Second Edition." In Isaac Newton, *Sir Isaac Newton's Mathematical Principles of Natural Philosophy and His System of the World*, edited by R. T. Crawford, xx–xxxiii. Translated by Andrew Motte (1729) and Florian Cajoli (1934). Vol. 1, *The Motion of Bodies*. New York: Greenwood Press, 1962.

Cottegnies, Line. "The Translator as Critic: Aphra Behn's Translation of Fontenelle's *Discovery of New Worlds* (1688)." *Restoration* 27, no. 1 (Spring 2003): 23–38.

Culler, Jonathan. "Omniscience." *Narrative* 12, no. 1 (January 2004): 22–34.

Dalmiya, Vrinda and Linda Alcoff. "Are 'Old Wives' Tales' Justified?" In Alcoff and Potter, *Feminist Epistemologies*, 217–44.

Daston, Lorraine and Peter Galison. *Objectivity*. Paperback edition. New York: Zone Books, 2010.

Davies, Tony. "The Ark in Flames: Science, Language and Education in Seventeenth-Century England." In *The Figural and the Literal: Problems of Language in the History of Science and Philosophy, 1630–1800*, edited by Andrew E. Benjamin, Geoffrey N. Cantor, and John R. R. Christie, 83–102. Manchester, UK: Manchester University Press, 1987.

Davis, Lennard. *Factual Fictions*. New York: Columbia University Press, 1983.

Davys, Mary. *The Accomplish'd Rake*. In Bowden, "*The Reform'd Coquet*," 127–226.

———. *The Amours of Alcippus and Lucippe. A Novel*. London: James Round, 1704.

———. *The Cousins; A Novel*. In *The Works of Mrs Davys*, 204–61. Vol. 2. London: H. Woodfall, 1725.

———. *Familiar Letters Betwixt a Gentleman and a Lady*. In Bowden, "*The Reform'd Coquet*," 93–120.

———. *The Fugitive*. London: W. Sawbridge, 1705.

———. *The Lady's Tale*. In *The Works of Mrs. Davys*, 123–201. Vol. 2. London: H. Woodfall, 1725.

———. *The Merry Wanderer*. In *The Works of Mrs. Davys*, 161–272. Vol. 1. London: H. Woodfall, 1725.

———. *The Reform'd Coquet*. In Bowden, "*The Reform'd Coquet*," 11–84.

Dawson, Hannah. *Locke, Language and Early-Modern Philosophy*. Cambridge: Cambridge University Press, 2007.

Dawson, Paul. "The Return of Omniscience in Contemporary Fiction." *Narrative* 17, no. 2 (May 2009): 144–61. doi:10.1353/nar.0.0023.

Dear, Peter. *Discipline and Experience: The Mathematical Way in the Scientific Revolution*. Chicago: University of Chicago Press, 1995.

———. "From Truth to Disinterestedness in the Seventeenth Century." *Social Studies of Science* 22, no. 4 (November 1992): 619–31. http://www.jstor.org/stable/285457.

———. *The Intelligibility of Nature: How Science Makes Sense of the World*. Chicago: University of Chicago Press, 2006.

———. "Totius in Verba: Rhetoric and Authority in the Early Royal Society." *Isis* 76, no. 2 (June 1985): 149–61.

De Bruyn, Frans. "Mary Davys (1674–1732)." In *Dictionary of Literary Biography: British Novelists, 1660–1800*, edited by Philip Breed Dematteis and Leemon B. McHenry, 131–38. Vol. 39. Detroit: Gale Research, 1985.

De Krey, Gary S. "Radicals, Reformers and Republicans: Academic Language and Political Discourse in Restoration London." In Houston and Pincus, *A Nation Transformed*, 71–99.

Descartes, René. *Discourse on Method*. In *Discourse on Method and Meditations on First Philosophy*. Translated by Donald A. Cress, 1–44. 4th ed. Indianapolis: Hackett, 1998.

Dictionary of National Biography. Edited by Leslie Stephen. Vol. 8, 1886. London: Smith, Elder, 1886. Hathi Trust Digital Library.

Donawerth, Jane. "Conversation and the Boundaries of Public Discourse in Rhetorical Theory by Renaissance Women." *Rhetorica: A Journal of the History of Rhetoric* 16, no. 2 (Spring 1998): 181–99. http://www.jstor.org/stable/10.1525/rh.1998.16.2.181.

Donne, John. "An Anatomy of the World: The First Anniversary." In *John Donne: The Complete Poems,* edited by A. J. Smith, 269–83. New York: Penguin, 1971.

Donovan, Josephine. "Women and the Framed-Novelle: A Tradition of Their Own." *Signs* 22, no. 4 (Summer 1997): 947–80.

———. *Women and the Rise of the Novel, 1405–1726.* New York: St. Martin's Press, 1999.

Dowd, Michelle M. and Julie A. Eckerle, eds. *Genre and Women's Life Writing in Early Modern England.* Burlington: Ashgate, 2007.

———. Introduction to Dowd and Eckerle, *Genre and Women's Life Writing,* 1–13.

Downie, J. A. "Mary Davys's 'Probable Feign'd Stories' and Critical Shibboleths about 'The Rise of the Novel.'" *Eighteenth-Century Fiction* 12, no. 2–3 (January–April 2000): 309–26. doi:10.1353/ecf.2000.0033.

Drury, Joseph. "Haywood's Thinking Machines." *Eighteenth-Century Fiction* 21, no. 2 (Winter 2008–9): 201–28.

Duffy, Maureen. Introduction to *Love-Letters Between a Nobleman and His Sister,* by Aphra Behn. Edited by Maureen Duffy, v–xvii. New York: Penguin, 1987.

———. *The Passionate Shepherdess: Aphra Behn, 1640–89.* London: Jonathan Cape, 1977.

Edmundson, Henry. *Lingua Linguarum….* London: T. Roycroft, 1655.

Eicke, Leigh A. "Jane Barker's Jacobite Writings." *Women's Writing and the Circulation of Ideas: Manuscript Publication in England, 1550–1800,* edited by George L. Justice and Nathan Tinker, 137–57. Cambridge: Cambridge University Press, 2002.

Espinosa, Aurelio M. "Hispanic Versions of the Tale of the Corpse Many Times 'Killed.'" *The Journal of American Folklore* 49, no. 193 (July–September 1936): 182–93.

Feingold, Mordechai. *The Newtonian Moment: Isaac Newton and the Making of Modern Culture.* New York: The New York Public Library and Oxford University Press, 2004.

Felton, Henry. *Resurrection of the Same Numerical Body….* London: Benjamin Motte, 1725.

Figueroa-Dorrego, Jorge. "Reconciling 'the most Contrary and Distant Thoughts': Paradox and Irony in the Novels of Aphra Behn." In *Re-Shaping the Genres: Restoration Women Writers,* edited by Zenón Luis-Martinez and Jorge Figueroa-Dorrego, 233–60. New York: Peter Lange, 2003.

Finger, Stanley. "The Lady and the Eel: How Aphra Behn Introduced Europeans to the 'Numb Eel.'" *Perspectives in Biology and Medicine 55*, no. 3 (Summer 2012): 378–401.

Fitzmaurice, James. "Daring and Innocence in the Poetry of Elizabeth Rochester and Jane Barker." *In-Between: Essays & Studies in Literary Criticism* 11, no. 1 (March 2002): 25–43.

———. "Jane Barker and the Tree of Knowledge at Cambridge University." *Renaissance Forum* 3, no. 1 (Spring 1998): np. http://www.hull.ac.uk /renforum/v3no1/fitzmaur.htm.

Fleming, James Dougal, ed. *The Invention of Discovery, 1500–1700.* Burlington: Ashgate, 2011.

Forstrom, J. Joanna S. *John Locke and Personal Identity: Immortality and Bodily Resurrection in 17th-Century Philosophy.* New York: Continuum, 2010.

Foucault, Michel. *The Archaeology of Knowledge & The Discourse on Language.* Translated by A. M. Sheridan Smith. New York: Pantheon, 1972.

Fox, Christopher. *Locke and the Scriblerians: Identity and Consciousness in Early Eighteenth-Century Britain.* Berkeley: University of California Press, 1988.

Frangos, Jennifer. "Ghosts in the Machine: The Apparition of Mrs. Veal, Rowe's *Friendship in Death* and the Early Eighteenth-Century Invisible World." In *Spirits Unseen: The Representation of Subtle Bodies in Early Modern European Culture*, edited by Christine Göttler and Wolfgang Neuber, 313–29. Boston: Brill, 2008.

Funkenstein, Amos. *Theology and the Scientific Imagination from the Middle Ages to the Seventeenth Century.* Princeton: Princeton University Press, 1986.

Gallagher, Catherine. *Nobody's Story: The Vanishing Acts of Women Writers in the Marketplace, 1670–1820.* Berkeley: University of California Press, 1994.

Genette, Gérard. *Figures of Literary Discourse.* Translated by Alan Sheridan. New York: Columbia University Press, 1982.

———. *Narrative Discourse: An Essay in Method.* Translated by Jane E. Lewin. Ithaca, NY: Cornell University Press, 1980.

Genovese, Elizabeth Fox. *Feminism Without Illusions: A Critique of Individualism.* Chapel Hill: University of North Carolina Press, 1991.

Gerrard, Christine. *Aaron Hill: The Muses' Projector, 1685–1750.* Oxford: Oxford University Press, 2003.

Gevirtz, Karen. "Aphra Behn and the Scientific Self." In *The New Science and Women's Literary Discourse: Prefiguring Frankenstein*, edited by Judy Hayden, 85–98. New York: Palgrave Macmillan, 2011.

———. *Life After Death: Widows and the English Novel, Defoe to Austen.* Newark, DE: University of Delaware Press, 2005.

Gilbert, Ruth. "Seeing and Knowing: Science, Pornography and Early Modern Hermaphrodites." In *At the Borders of the Human: Beasts,*

Bodies and Natural Philosophy in the Early Modern Period, edited by Erica Fudge, Ruth Gilbert, and Susan Wiseman, 150–70. New York: St. Martin's Press, 1999.

Gilroy, Amanda and W. M. Verhoeven. Introduction to Gilroy and Verhoeven, *Epistolary Histories*, 1–25.

Gilroy, Amanda and W. M. Verhoeven, eds. *Epistolary Histories*. Charlottesville: University Press of Virginia, 2000.

Girten, Kristen M. "Unsexed Souls: Natural Philosophy as Transformation in Eliza Haywood's *Female Spectator*." *Eighteenth-Century Studies* 43, no. 1 (Fall 2009): 55–74.

Glanvill, Joseph. *Scepsis Scientifica: OR, Confest Ignorance, the way to Science; In an Essay of The Vanity of Dogmatizing, and Confident Opinion, with a Reply to the Exceptions Of the Learned Thomas Albius.* London: E. Cotes, 1665.

Gleick, James. *Isaac Newton.* New York: Vintage Books, 2003.

Glover, Susan. *Engendering Legitimacy: Law, Property, and Early Eighteenth-Century Fiction.* Lewisburg: Bucknell University Press, 2006.

Golinski, Jan. *British Weather and the Climate of Enlightenment.* Chicago: University of Chicago Press, 2011. eBook.

———. "Robert Boyle: Skepticism and Authority in Seventeenth-Century Chemical Discourse." In *The Figural and the Literal: Problems of Language in the History of Science and Philosophy, 1630–1800*, edited by Andrew E. Benjamin, Geoffrey N. Cantor, and John R. R. Christie, 58–82. Manchester, UK: Manchester University Press, 1987.

———. *Science as Public Culture: Chemistry and Enlightenment in Britain, 1760–1820.* Cambridge: Cambridge University Press, 1992.

Goodfellow, Sarah. "'Such Masculine Strokes': Aphra Behn as Translator of *A Discovery of New Worlds*." *Albion: A Quarterly Journal Concerned with British Studies* 28, no. 2 (Summer 1996): 229–50.

Goreau, Angeline. *Reconstructing Aphra: A Social Biography of Aphra Behn.* New York: Dial, 1980.

Gorham, Geoffrey. "Mind-Body Dualism and the Harvey-Descartes Controversy." *Journal of the History of Ideas* 55, no. 2 (April 1994): 211–34. doi:10.2307/2709897.

Göttler, Christine. Preface: "Vapours and Veils: The Edge of the Unseen." In *Spirits Unseen: The Representation of Subtle Bodies in Early Modern European Culture*, edited by Christine Göttler and Wolfgang Neuber, xv–xxvii. Boston: Brill, 2008.

Gowing, Laura. *Common Bodies: Women, Touch and Power in Seventeenth-Century England.* New Haven: Yale University Press, 2003.

Graham, Elspeth. "Intersubjectivity, Intertextuality, and Form in the Self-Writings of Margaret Cavendish." In Dowd and Eckerle, *Genre and Women's Life Writing*, 131–50.

Greene, Richard. *The Principles of the Philosophy of the Expansive and Contractive Forces. Or an Inquiry into the Principles of the Modern Philosophy....* Cambridge: Cornelius Crownfield, 1727.

Gretton, Phillips. *A Review of the Argument a priori, In Relation to the Being and Attributes of God....* London: Bernard Lintot, 1726.

Grosz, Elizabeth. "Bodies and Knowledges: Feminism and the Crisis of Reason." In Alcoff and Potter, *Feminist Epistemologies*, 187–215.

Gunn, Daniel P. "Free Indirect Discourse and Narrative Authority in *Emma*." *Narrative* 12, no. 1 (January 2004): 35–54.

Hall, Rupert. "Isaac Newton: Creator of the Cambridge Scientific Tradition." In *Cambridge Scientific Minds*, edited by Peter Harman and Simon Mitton, 36–50. Cambridge: Cambridge University Press, 2002.

Halley, Edmund. "On this Work of Mathematical Physics Of the Most Outstanding Man Most Learned Isaac Newton." In Newton, *Principia Naturalis*, 233–34.

Hammond, Brean and Shaun Regan. *Making the Novel: Fiction and Society in Britain, 1660–1789*. New York: Palgrave Macmillan, 2006.

Handley, Sasha. *Visions of an Unseen World: Ghost Beliefs and Ghost Stories in Eighteenth-Century England*. London: Pickering & Chatto, 2007.

Hannay, Margaret P. "'How I These Studies Prize': The Countess of Pembroke and Elizabethan Science." In Hunter and Hutton, *Women, Science and Medicine*, 108–21.

Haraway, Donna. "Modest_Witness@Second_Millennium." In *The Haraway Reader*, 223–50. New York: Routledge, 2004.

Harris, Frances. "Living in the Neighbourhood of Science: Mary Evelyn, Margaret Cavendish and the Greshamites." In Hunter and Hutton, *Women, Science and Medicine*, 198–217.

Harrison, Peter. *The Bible, Protestantism, and the Rise of Natural Science*. New York: Cambridge University Press, 1998.

———. "Curiosity, Forbidden Knowledge, and the Reformation in Natural Philosophy in Early Modern England." *Isis* 92, no. 2 (June 2001): 265–90. http://www.jstor.org/stable/3080629.

———. *The Fall of Man and the Foundations of Science*. New York: Cambridge University Press, 2007.

———. "Newtonian Science, Miracles, and the Laws of Nature." *Journal of the History of Ideas* 56, no. 4 (October 1995): 531–53. http://www.jstor.org/stable/2709991.

Harrow, Sharon. "Having Text: Desire and Language in Haywood's *Love in Excess* and *The Distressed Orphan*." *Eighteenth-Century Fiction* 22, no. 2 (Winter 2009–10): 279–308.

Harth, Erica. *Cartesian Women: Versions and Subversions of Rational Discourse in the Old Regime*. Ithaca, NY: Cornell University Press, 1992.

Harvey, Karen. *Reading Sex in the Eighteenth Century: Bodies and Gender in English Erotic Culture*. Cambridge: Cambridge University Press, 2004.

———. "Refinement in a Teacup? Punch, Domesticity and Gender in the Eighteenth Century." *Journal of Design History* 21, no. 3 (Autumn 2008): 205–21. doi:10.1093/jdh/epn022.

———. "The Substance of Sexual Difference: Change and Persistence in Representations of the Body in Eighteenth-Century England." *Gender & History* 14, no. 2 (August 2002): 202–23.

Harvey, William. *On the Motion of the Heart and Blood in Animals.* Translated by Robert Willis. Buffalo: Prometheus, 1993.

Harwood, John T. "Rhetoric and Graphics in *Micrographia.*" In *Robert Hooke: New Studies,* edited by Michael Hunter and Simon Schaffer, 119–47. Woodbridge, UK: Boydell, 1989.

Hayden, Judy A., ed. *The New Science and Women's Literary Discourse: Prefiguring Frankenstein.* New York: Palgrave Macmillan, 2011.

———. "Women, Education, and the Margins of Science." In Hayden, *The New Science,* 1–15.

Haywood, Eliza. "The British Recluse." In *Popular Fiction by Women 1660–1730: An Anthology,* edited by Paula R. Backscheider and John J. Richetti, 153–224. Oxford: Oxford University Press, 1996.

———. *The Female Spectator.* In *Selections from "The Female Spectator,"* edited by Patricia Meyers Spacks. New York: Oxford University Press, 1999.

———. *Love in Excess.* Edited by David Oakleaf, 37–266. 2nd ed. Toronto: Broadview, 2000.

———. *The Tea-Table: Or, A Conversation between Some Polite Persons of Both Sexes.* In *"Fantomina" and Other Works,* edited by Alexander Pettit, Margaret Case Croskery, and Anna C. Patchias, 73–106. Toronto: Broadview Press, 2004.

———. *The Tea-Table: Or, A Conversation between some Polite Persons of both Sexes...Part the Second.* In *Selected Works of Eliza Haywood,* edited by Alexander Pettit, 35–59. Vol. 1, *Miscellaneous Writings, 1725–43.* London: Pickering & Chatto, 2000.

Heinze, Reudiger. "Violations of Mimetic Epistemology in First-Person Narrative Fiction." *Narrative* 16, no. 3 (October 2008): 279–97.

Henry, John. "Religion and the Scientific Revolution." In *The Cambridge Companion to Science and Religion,* edited by Peter Harrison, 39–58. New York: Cambridge University Press, 2010.

Herman, David. "Narrative Theory after the Second Cognitive Revolution." In *Introduction to Cognitive Cultural Studies,* edited by Lisa Zunshine, 155–75. Baltimore: The Johns Hopkins University Press, 2010.

Hill, Aaron. *An Account of the Rise and Progress of the Beech-Oil Invention, and All the Steps Which Have Been Taken in That Affair, from the First Discovery, to the Present Time....* London, 1715.

———. *The Plain Dealer: Being Select Essays on Several Curious Subjects, Relating to Friendship....* Vol. 1. London, 1730.

Hnath, Lucas. *Isaac's Eye.* Dir. Linsay Firman, perf. Jeff Biehl, Kristen Bush, Haskell King and Michael Louis Serafin-Wells. Ensemble Studio Theatre, New York, 2013.

Hogan, Patrick Colm. *The Mind and Its Stories: Narrative Universals and Human Emotion.* Cambridge: Cambridge University Press, 2003.

Honeybone, Michael. "Sociability, Utility and Curiosity in the Spalding Gentlemen's Society, 1710–60." In *Science and Beliefs: From Natural Philosophy to Natural Science, 1700–1900*, edited by David M. Knight and Matty D. Eddy, 63–76. Aldershot: Ashgate, 2005.

Hooke, Robert. *Micrographia*. London: Jo. Martyn and Ja. Allestry, 1665. http://www.gutenberg.org/files/15491/15491-h/15491-h.htm.

Houston, Alan and Steve Pincus, eds. *A Nation Transformed: England After the Restoration*. Cambridge: Cambridge University Press, 2001.

Hultquist, Aleksondra. "Marriage in Haywood; or, Amatory Reading Rewarded." In Carlile, *Masters*, 31–46.

Hunter, J. Paul. *Before Novels: The Cultural Contexts of Eighteenth-Century English Fiction*. New York: W. W. Norton, 1990.

———. "Robert Boyle and the Epistemology of the Novel." *Eighteenth-Century Fiction* 2, no. 4 (1990): 275–91.

Hunter, Lynette. "Sisters of the Royal Society: The Circle of Katherine Jones, Lady Ranelagh," in Hunter and Hutton, *Women, Science and Medicine*, 178–97.

Hunter, Lynette and Sarah Hutton. Introduction to Huntter and Hutton, *Women, Science and Medicine*, 1–6.

———, eds., *Women, Science and Medicine 1500–1700: Mothers and Sisters of the Royal Society*. Stroud: Sutton Publishing, 1997.

Hunter, Michael. *Boyle: Between God and Science*. New Haven: Yale University Press, 2009.

———. "The Debate over Science." In Hunter, *Science and the Shape of Orthodoxy*, 101–19.

———. *Establishing the New Science: The Experience of the Royal Society*. Woodbridge, UK: Boydell, 1989.

———. "Introduction: Fifteen Essays and a New Theory of Intellectual Change." In Hunter, *Science and the Shape of Orthodoxy*, 1–18.

———. "The Making of Christopher Wren." In Hunter, *Science and the Shape of Orthodoxy*, 45–65.

———. "Science and Astrology in Seventeenth-Century England: An Unpublished Polemic by John Flamsteed." In Hunter, *Science and the Shape of Orthodoxy*, 245–85.

———. *Science and the Shape of Orthodoxy: Intellectual Change in Late Seventeenth-Century Britain*. Woodbridge, UK: Boydell, 1995.

———. *Science and Society in Restoration England*. Aldershot: Ashgate, 1992.

Hutchinson, Keith. "What Happened to Occult Qualities in the Scientific Revolution?" *Isis* 73, no. 2 (June 1982): 233–53. http://www.jstor.org/stable/231676.

Hutton, Sarah. "The Riddle of the Sphinx: Francis Bacon and the Emblems of Science." In Hunter and Hutton, *Women, Science and Medicine*, 7–28.

———. "Sisters of the Royal Society: The Circle of Katherine Jones, Lady Ranelagh." In Hunter and Hutton, *Women, Science and Medicine*, 178–97.

Iliffe, Roger. "Author-Mongering: The 'Editor' Between Producer and Consumer." In *The Consumption of Culture 1600–1800: Image, Object, Text*, edited by Ann Bermingham and John Brewer, 166–92. New York: Routledge, 1995.

———. *Newton: A Very Short Introduction*. New York: Oxford University Press, 2007. eBook.

Johnson, George. *The Ten Most Beautiful Experiments*. New York: Knopf, 2008.

Johnson, Samuel. *A Dictionary of the English Language....* Vol. 1. London: J. Knapton; C. Hitch and L. Hawes; A. Millar, R. and J. Dodsley; and M. and T. Longman, 1766.

Joule, Victoria. "Mary Davys's Novel Contribution to Women and Realism." *Women's Writing* 17, no. 1 (May 2010): 30–48. doi:10.1080/09699080903533262.

Katritzky, M. A. *Women, Medicine and Theatre, 1500–1750*. Burlington: Ashgate, 2007.

Keeran, Peggy and Jennifer Bowers. *Literary Research and the British Eighteenth Century: Strategies and Sources*. Lanham, MD: Scarecrow Press, 2013.

Keller, Eve. "Producing Petty Gods: Margaret Cavendish's Critique of Experimental Science." *ELH* 64, no. 2 (1997): 447–71. doi:10.1353/elh.1997.0017.

Kern, Jean B. "Mary Davys as Novelist of Manners." *Essays in Literature* 10, no. 1 (Spring 1983): 29–38.

King, Kathryn R. *Jane Barker, Exile: A Literary Career, 1675–1725*. Oxford: Clarendon Press, 2000.

———. "Jane Barker, *Poetical Recreations*, and the Sociable Text." *ELH* 61, no 3 (Autumn 1994): 551–70. http://www.jstor.org/stable/2873334.

———. "Of Needles and Pens and Women's Work." *Tulsa Studies in Women's Literature* 14, no. 1 (Spring 1995): 77–93.

———. "Patriot or Opportunist? Eliza Haywood and the Politics of *The Female Spectator*." In Wright and Newman, *Fair Philosopher*, 104–21.

———. *A Political Biography of Eliza Haywood*. London: Pickering & Chatto, 2012.

———. "Spying Upon the Conjurer: Haywood, Curiosity, and 'the Novel' in the 1720s." *Studies in the Novel* 30, no. 2 (Summer 1998): 178–93.

King, Kathryn R. and Jesslyn Medoff. "Jane Barker and Her Life (1652–1732): The Documentary Record." *Eighteenth-Century Life* 21 (November 1997): 16–38. http://muse.jhu.edu/journals/eighteenth-century_life/v021/21.3king.html.

Kramnick, Jonathan. *Actions and Objects from Hobbes to Richardson*. Stanford: Stanford University Press, 2010.

Kronk, Gary W. *Cometology: A Catalog of Comets*. Vol. 1: *Ancient-1799*. Cambridge: Cambridge University Press, 1999.

Lanser, Susan Sniader. *Fictions of Authority: Women Writers and Narrative Voice*. Ithaca, NY: Cornell University Press, 1992.

Lanser, Susan Sniader. *The Narrative Act: Point of View in Prose Fiction.* Princeton: Princeton University Press, 1981.

Lamarra, Annamaria. "The Difficulty in Saying 'I': Aphra Behn and the Female Autobiography." In *Aphra Behn (1640–1689): Le Modèle Européen*, edited by Mary Ann O'Donnell and Bernard Dhuicq, 1–7. Entrevaux, France: Bilingua GA Editions, 2005.

Laqueur, Thomas. *Making Sex: Body and Gender from the Greeks to Freud.* Cambridge, MA: Harvard University Press, 1990.

Larsen, Kristine. "'A Woman's Place is in the Dome': Gender and the Astronomical Observatory, 1670–1970." *MP: An Online Feminist Journal* 2, no. 5 (October 2009): 104–24. http://academinist.org/wp-content /uploads/2009/10/Woman_Place_Larsen.pdf.

Laslett, Peter. Introduction to *Two Treatises of Government*, edited by Peter Laslett, 3–122. Student Edition. Cambridge: Cambridge University Press, 1988.

Leary, Jr., John E. *Francis Bacon and the Politics of Science.* Ames: Iowa State University Press, 1994.

van Leeuwenhoek, Antonie. "Microscopical Observations of Mr. Leewenhoeck, Concerning the Optic Nerve, Communicated to the Publisher in Dutch, and by Him Made English," part 1. *Philosophical Transactions* 10 (1675): 379. http://www.jstor.org/stable/101663.

Lehrer, Jonah. *Proust Was a Neuroscientist.* New York: Houghton Mifflin, 2004.

Lewis, Jayne Elizabeth. *Air's Appearance: Literary Atmosphere in British Fiction, 1660–1794.* Chicago: University of Chicago Press, 2012.

Locke, John. *An Essay Concerning Human Understanding.* 6th ed. Kitchener, ON: Batoche Books, 2001. eBook.

Lodowyck, Francis. *The Ground-Work, or Foundation, Laid (or So Intended) for the Framing of a New Perfect Language....* London: 1652.

Lovell, Terry. "Subjective Powers?: Consumption, the Reading Public, and Domestic Woman in Early Eighteenth-Century England." In *The Consumption of Culture 1600–1800: Image, Object, Text*, edited by Ann Bermingham and John Brewer, 22–41. New York: Routledge, 1995.

Lubey, Kathleen. "Eliza Haywood's Amatory Aesthetic." *Eighteenth-Century Studies* 39, no. 3 (2006): 309–22.

MacKenzie, Niall. "Jane Barker, Louise Hollandine of the Palatinate and 'Solomons Wise Daughter.'" *The Review of English Studies*, New Series 58, no. 233 (2007): 64–72. doi:10.1093/res/hgl142.

Maclaurin, Colin. *An Account of Sir Isaac Newton's Philosophical Discoveries....* London: A. Millar and J. Nourse, 1748.

MacLean, Gerald. Postscript. In Gilroy and Verhoeven, *Epistolary Histories*, 170–72.

———."Re-Siting the Subject." In Gilroy and Verhoeven, *Epistolary Histories*, 176–97.

Mascuch, Michael. *Origins of the Individualist Self: Autobiography and Self-Identity in England, 1591–1791.* Stanford: Stanford University Press, 1996.

McArthur, Tonya Moutray. "Jane Barker and the Politics of Catholic Celibacy." *SEL Studies in English Literature 1500–1900* 47, no. 3 (Summer 2007): 595–618. doi:10.1353/sel.2007.0030.

McCann, Edwin. "Locke's Philosophy of Body." In *The Cambridge Companion to Locke*, edited by Vere Chappell, 56–60. Cambridge: Cambridge University Press, 1994.

McBurney, William H. "Mrs. Mary Davys: Forerunner of Fielding." *PMLA* 74, no. 4 (September 1959): 348–55. http://www.jstor.org/stable /460444

McKeon, Michael. *The Origins of the English Novel, 1600–1740*. Baltimore: The Johns Hopkins University Press, 1987.

Meadows, A. J. "John Flamsteed, Our Astronomical Observator." *Notes and Records of the Royal Society of London*, 50, no. 2 (July 1996): 250–53. http://www.jstor.org/stable/531915.

Merritt, Juliette. *Beyond Spectacle: Eliza Haywood's Female Spectators*. Toronto: University of Toronto Press, 2006.

———. "Reforming the Coquet? Eliza Haywood's Vision of a Female Epistemology." In Wright and Newman, *Fair Philosopher*, 176–92.

Messenger, Ann. *His and Hers: Essays in Restoration and Eighteenth-Century Literature*. Lexington, KY: The University Press of Kentucky, 1986.

Moglin, Helene. *The Trauma of Gender: A Feminist Theory of the English Novel*. Berkeley: University of California Press, 2001.

Molesworth, Jesse. *Chance and the Eighteenth-Century Novel: Realism, Probability, Magic*. New York: Cambridge University Press, 2010.

Morgan, Thomas. *Philosophical Principles of Medicine....* London: J. Darby and T. Browne, 1725.

"A Narrative Concerning the Success of Pendulum-Watches at Sea for the Longitudes." *Philosophical Transactions* 1, no. 1 (1665–1666): 13–14. http://www.jstor.org/stable/101409.

Nelson, Lynn Hankinson. "Epistemological Communities." In Alcoff and Potter, *Feminist Epistemologies*, 121–59.

Newton, Isaac. Laboratory Notebook, c. 1669–1693, Newton Papers. Cambridge University Library, Cambridge, UK.

———. "Preface to the Reader." In *Key to Newton's Dynamics: The Kepler Problem and the Principia: Containing an English Translation of Sections 1, 2, and 3 of Book One From the First (1687) Edition of Newton's Mathematical Principles of Natural Philosophy*, 230–67. Translated by J. Bruce Brackenridge. Berkeley: University of California Press, 1995. eBook.

———. *Sir Isaac Newton's Mathematical Principles of Natural Philosophy and His System of the World*, edited by R. T. Crawford. Translated by Andrew Motte (1729) and Florian Cajoli (1934). Vol. 1, *The Motion of Bodies*. New York: Greenwood Press, 1962.

Nicolson, Marjorie Hope. *Newton Demands the Muse: Newton's "Opticks" and the Eighteenth-Century Poets*. Princeton: Princeton University Press, 1946.

Novak, Maximilian E. "Friday: or, the Power of Naming." In *Augustan Subjects: Essays in Honor of Martin C. Battestin*, edited by Albert J. Rivero, 110–22. Newark, DE: University of Delaware Press, 1997.

O'Donnell, Mary Ann. *Aphra Behn: An Annotated Bibliography of Primary and Secondary Sources*. 2nd ed. Burlington: Ashgate, 2004.

Orr, Leah. "Attribution Problems in the Fiction of Aphra Behn." *The Modern Language Review* 108, no. 1 (January 2013): 30–51. http://www.jstor.org/stable/10.5699/modelangrevi.108.1.0030.

———. "The Basis for Attribution in the Canon of Eliza Haywood." *The Library: The Transactions of the Bibliographical Society* 12, no. 4 (December 2011): 335–75. http://muse.jhu.edu/journals/lbt/summary/v012/12.4.orr.html.

Oxford English Dictionary online. Oxford: Oxford University Press, 2013.

Parker, J. *The History of Eclipses....* London: J. Read and R. Burleigh, 1715.

Pateman, Carol. *The Sexual Contract*. Stanford: Stanford University Press, 1988.

Pears, Iain. *An Instance of the Fingerpost*. New York: Berkley Books, 1998.

Pearson, Jacqueline. "Gender and Narrative in the Fiction of Aphra Behn." Pt. 1. *The Review of English Studies* 42, no. 165 (February 1991): 40–56.

Pechey, John. *The Compleat Herbal of Physical Plants....* London: R. and J. Bonwicke, 1707.

Perronet, Vincent. *A Vindication of Mr. Locke*. London: James, John, and Paul Knapton, 1736.

Perry, Ruth. *Women, Letters, and the Novel*. New York: AMS Press, 1980.

Pettit, Alexander. "Adventures in Pornographic Places: Eliza Haywood's *Tea-Table* and the Decentering of Moral Argument." *Papers on Language & Literature* 38, no. 3 (Summer 2002): 244–69.

Pettit, Alexander, Margaret Case Croskery, and Anna C. Patchias. Introduction to *"Fantomina" and Other Works*, edited by Alexander Pettit, Margaret Case Croskery, and Anna C. Patchias, 9–32. Toronto: Broadview, 2004.

Petty, William. *Another Essay in Political Arithmetick, Concerning the Growth of the City of London with the Measures, Periods, Causes, and Consequences Thereof, 1682*. London: H. H. for Mark Pardoe, 1683.

———. "An Extract of Two Essays in Political Arithmetick concerning the Comparative Magnitudes, etc. of London and Paris by Sr. William Petty Knight. R. S. S." *Philosophical Transactions* 16 (1686–1692): 152. http://www.jstor.org/stable/101856.

Pitt, Robert. *The Craft and Frauds of Physick Expos'd....* London: Tim Childe, 1702.

P. L. *Two Essays Sent in a Letter from Oxford, to a Nobleman in London...* London: R. Baldwin, 1695.

Poovey, Mary. *Genres of the Credit Economy*. Chicago: University of Chicago Press, 2008.

Porter, Roy. *Flesh in the Age of Reason*. New York: W. W. Norton, 2003.

Potter, Elizabeth. *Gender and Boyle's Law of Gases*. Bloomington: Indiana University Press, 2001.

———. "Gender and Epistemic Negotiation." In Alcoff and Potter, *Feminist Epistemologies*, 161–86.

Potter, Tiffany. "'A God-like Sublimity of Passion': Eliza Haywood's Libertine Consistency." *The Eighteenth-Century Novel*, edited by Susan Spencer, 95–126. Vol. 1. New York: AMS Press, 2001.

Power, Henry. *Experimental Philosophy, In Three Books: Containing New Experiments Microscopical, Mercurial, Magnetical....* London: T. Roycroft, 1664.

The Preface to *Philosophical Transactions*, 16 (1686–1692): 36. http://www.jstor.org/stable/101838.

Prescott, Sarah. *Women, Authorship and Literary Culture, 1690–1740*. New York: Palgrave Macmillan, 2003.

Pritchard, Will. *Outward Appearances: The Female Exterior in Restoration London*. Lewisburg: Bucknell University Press, 2008.

Reed, Walter L. *An Exemplary History of the Novel*. Chicago: University of Chicago Press, 1981.

Reiss, Timothy J. *Mirages of the Selfe: Patterns of Personhood in Ancient and Early Modern Europe*. Stanford: Stanford University Press, 2003.

R. F. *The Pure Language of the Spirit of Truth....* London: Giles Calvert, 1655.

Richardson, Alan. *The Neural Sublime: Cognitive Theories and Romantic Texts*. Baltimore: The Johns Hopkins University Press, 2010.

Richetti, John. "*Love Letters Between a Nobleman and His Sister*: Aphra Behn and Amatory Fiction." In *Augustan Subjects: Essays in Honor of Martin C. Battestin*, edited by Albert J. Rivero, 13–28. Newark, DE: University of Delaware Press, 1997.

Riley, Lindy. "Mary Davys's Satiric Novel *Familiar Letters*: Refusing Patriarchal Inscription of Women." In *Cutting Edges: Postmodern Critical Essays on Eighteenth-Century Satire*, edited by James E. Gill, 206–21. Knoxville: University of Tennessee Press, 1995.

Rivero, Albert J. "'Heiroglyphick'd' History: in Aphra Behn's *Love-Letters between a Nobleman and His Sister*." *Studies in the Novel* 30, no. 2 (Summer 1998): 126–38.

Rousseau, George S. *Enlightenment Borders: Pre- And Post-Modern Discourses: Medical, Scientific*. Manchester, UK: Manchester University Press, 1991.

The Royal Society. "Second Charter of the Royal Society for Improving Natural Knowledge (1663)." http://royalsociety.org/Charters-of-the-Royal-Society/.

———. "Statistics." http://royalsociety.org/about-us/equality/statistics/.

Royle, Nicholas. *The Uncanny*. New York: Routledge, 2003.

Rubik, Margarete. "Estranging the Familiar, Familiarizing the Strange: Self and Other in *Oroonoko* and *The Widdow Ranter*." In *Aphra*

Behn (1640–1689): Identity, Alterity, Ambiguity, edited by Mary Ann O'Donnell, Bernard Dhuicq, and Guyonne Leduc, 33–41. Montreal: L'Harmattan, 2000.

Salmon, William. *Pharmacopoia Londinensis. Or, the New London Dispensatory....* London: Royal College of Physicians, 1702.

Sargent, Rose-Mary, ed. *Francis Bacon: Selected Philosophical Works.* Indianapolis: Hackett, 1999.

Saxton, Kirsten T. Introduction to *The Passionate Fictions of Eliza Haywood: Essays on Her Life and Work*, edited by Kirsten T. Saxton and Rebecca P. Bocchicchio, 1–18. Lexington, KY: The University Press of Kentucky, 2000.

Saxton, Kirsten T. and Rebecca P. Bocchicchio, eds. *The Passionate Fictions of Eliza Haywood: Essays on Her Life and Work.* Lexington, KY: The University Press of Kentucky, 2000.

Schiebinger, Londa. *The Mind Has No Sex? Women in the Origins of Modern Science.* Cambridge, MA: Harvard University Press, 1989.

Schneider, Gary. *The Culture of Epistolarity: Vernacular Letters and Letter Writing in Early Modern England, 1500–1700.* Newark, DE: University of Delaware Press, 2005.

Scott, Joan Wallach. *Gender and the Politics of History.* Rev. Ed. New York: Columbia University Press, 1999. eBook.

Shapin, Steven. *Never Pure: Historical Studies of Science as If It Was Produced by People with Bodies, Situated in Time, Space, Culture, and Society, and Struggling for Credibility and Authority.* Baltimore: The Johns Hopkins University Press, 2010.

———. *The Scientific Revolution.* Chicago: University of Chicago Press, 1996.

———. *A Social History of Truth.* Chicago: University of Chicago Press, 1994.

Shapin, Steven and Simon Shaffer. *Leviathan and the Air Pump: Hobbes, Boyle, and the Experimental Life. Including and Translation of Thomas Hobbes, "Dialogus Physicus de Natura Aeris," by Simon Shaffer.* Princeton: Princeton University Press, 1985.

Shapiro, Barbara. "History and Natural History in Sixteenth- and Seventeenth-Century England: An Essay on the Relationship between Humanism and Science." In *English Scientific Virtuosi in the 16th and 17th Centuries*, papers read at a Clark Library Seminar, February 3, 1977, edited by Barbara Shapiro and Robert G. Frank, Jr., 3–55. Los Angeles: William Andrews Clark Memorial Library, 1977.

———. "Natural Philosophy and Political Periodization: Interregnum, Restoration and Revolution." In Houston and Pincus, *A Nation Transformed*, 299–327.

Shaw, Narelle. "Free Indirect Speech and Jane Austen's 1816 Revision of *Northanger Abbey.*" *Studies in English Literature, 1500–1900* 30, no. 4 (Autumn 1990): 591–601. http://www.jstor.org/stable/450561.

Sheridan, Patricia. Introduction to *Catharine Trotter Cockburn: Philosophical Writings*, edited by Patricia Sheridan, 17–19. Toronto: Broadview, 2006.

Shevelow, Kathryn. *Women and Print Culture: The Construction of Femininity in the Early Periodical*. New York: Routledge: 1989.

Slare, Frederick. *Experiments and Observations upon Oriental and Other Bezoar-Stones, Which Prove Them to Be of No Use in Physic....* London: Timothy Goodwin, 1715.

Smith, Amy Elizabeth. "Naming the Un-'Familiar': Formal Letters and Travel Narratives in Late Seventeenth- and Eighteenth-Century Britain." *The Review of English Studies* 54, no. 214 (May 2003): 178–202. http://res.oxfordjournals.org.ezproxy.shu.edu/content/54/214/178.

Smith, George. *A Compleat Body of Distilling, Explaining the Mysteries of That Science, in a Most Easy and Familiar Manner....* London: Bernard Lintot, 1725.

Smith, Justin E. H. "Spirit as Intermediary in Post-Cartesian Natural Philosophy." In *Spirits Unseen: The Representation of Subtle Bodies in Early Modern European Culture*, edited by Christine Göttler and Wolfgang Neuber, 269–91. Boston: Brill, 2008.

Snider, Alvin. "Atoms and Seeds: Aphra Behn's Lucretius." *Clio* 33, no. 1 (Fall 2003): 1–24.

———. "Cartesian Bodies." *Modern Philology: A Journal Devoted to Research in Medieval and Modern Literature* 92, no. 2 (November 2000): 299–319.

Sobel, Dava. *Longitude: The True Story of a Lone Genius Who Solved the Greatest Scientific Problem of His Time*. New York: Walker and Company, 1995.

Spedding, Patrick. *A Bibliography of Eiza Haywood*. London: Pickering & Chatto, 2004.

Spencer, Jane. "Amatory and Scandal Fiction." In *The Oxford History of the English Novel*, edited by Thomas Keymer. Vol. 1. Oxford: Oxford University Press, forthcoming.

———. "Creating the Woman Writer: The Autobiographical Works of Jane Barker." *Tulsa Studies in Women's Literature* 2, no. 2 (Autumn 1983): 165–81.

———. *The Rise of the Woman Novelist: From Aphra Behn to Jane Austen*. Oxford: Blackwell, 1986.

Spolsky, Ellen. "Narrative as Nourishment." In *Toward a Cognitive Theory of Narrative Acts*, edited by Frederick Luis Aldama, 37–60. Austin: University of Texas Press, 2010.

Sprat, Thomas. *History of the Royal Society*, edited by Jackson I. Cope and Harold Whitmore Jones. St. Louis: Washington University Press, 1958.

Staves, Susan. *A Literary History of Women's Writing in Britain, 1660–1789*. New York: Cambridge University Press, 2008.

Steen, Francis F. "The Politics of Love: Propaganda and Structural Learning in Aphra Behn's *Love-Letters between a Nobleman and His Sister.*" *Poetics Today* 23, no. 1 (Spring 2002): 91–122.

Stefanson, Donald Hal. "The Works of Mary Davys: A Critical Edition (Vol. I and II)." PhD diss., University of Iowa, 1971.

Strawson, Galen. *Locke on Personal Identity: Consciousness and Concernment.* Princeton: Princeton University Press, 2011.

Stubbe, Henry. *The Indian Nectar, or, A Discourse Concerning Chocolate....* London: J. C. for Andrew Cook, 1662.

Swenson, Rivka. "Representing Modernity of Jane Barker's *Galesia Trilogy*: Jacobite Allegory and the Patch-Work Aesthetic." *Studies in Eighteenth-Century Culture* 34 (2005): 55–80.

Taylor, Charles. *Sources of the Self: The Making of the Modern Identity.* Cambridge, MA: Harvard University Press, 1989.

Thompson, Helen. "Plotting Materialism: W. Charleton's *The Ephesian Matron*, E. Haywood's *Fantomina*, and Feminine Consistency." *Eighteenth-Century Studies* 35, no. 2 (Winter 2002): 195–214. doi:10.1353/ecs.2002.0017.

———. "'Thou Monarch of my Panting Soul': Hobbesian Obligation and the Durability of Romance in Aphra Behn's *Love-Letters.*" In *British Women's Writing in the Long Eighteenth Century: Authorship, Politics and History*, edited by Jennie Batchelor and Cora Kaplan, 107–20. New York: Palgrave Macmillan, 2005.

Thompson, Helen and Natania Meeker. "Empiricism, Substance, Narrative: An Introduction." Special issue, *Eighteenth Century: Theory and Interpretation* 48, no. 3 (Fall 2007): 183–86.

Tierney-Hynes, Rebecca. *Novel Minds: Philosophers and Romance Readers, 1680–1740.* New York: Palgrave Macmillan, 2012.

Todd, Janet. General Introduction to *The Works of Aphra Behn*, edited by Janet Todd, ix–xxxv. Vol. 1: *Poetry.* Columbus: Ohio State University Press, 1992.

———. "*Love-Letters* and Critical History." In *Aphra Behn (1640–1689): Identity, Alterity, Ambiguity*, edited by Mary Ann O'Donnell, Bernard Dhuicq, and Guyonne Leduc, 197–201. Montreal: L'Harmattan, 2000.

———. *The Secret Life of Aphra Behn.* London: Pandora, 2000.

———. *The Sign of Angellica: Women, Writing and Fiction, 1660–1800.* New York: Columbia University Press, 1989.

———. Textual introduction to *The Works of Aphra Behn*, edited by Janet Todd, xxxvi–xlviii. Vol. 1: *Poetry.* Columbus: Ohio State University Press, 1992.

———. Textual introduction to *The Works of Aphra Behn*, edited by Janet Todd, vii–xviii. Vol. 3: *"The Fair Jilt" and Other Stories.* Columbus: Ohio State University Press, 1995.

———. Textual introduction to *The Works of Aphra Behn*, edited by Janet Todd, vii–xviii. Vol. 4: *"Seneca Unmasqued" and Other Prose Translations.* Columbus: Ohio State University Press, 1993.

Turner, Martha A. *Mechanism and the Novel: Science in the Narrative Process*. Cambridge: Cambridge University Press, 1993.

Vermeule, Blakey. *The Party of Humanity: Writing Moral Psychology in Eighteenth-Century Britain*. Baltimore: The Johns Hopkins University Press, 2000.

Wahrman, Dror. *The Making of the Modern Self: Identity and Culture in Eighteenth-Century England*. New Haven: Yale University Press, 2004.

Wakely, Alice. "Mary Davys and the Politics of Epistolary Form." In *"Cultures of Whiggism": New Essays on English Literature and Culture in the Long Eighteenth Century*, edited by David Womersley, assisted by Paddy Bullard and Abigail Williams, 257–67. Newark, DE: University of Delaware Press, 2005.

Wallwork, Jo. "Disruptive Behaviour in the Making of Science: Cavendish and the Community of Seventeenth-Century Science." In Wallwork and Salzman, *Early Modern Englishwomen*, 4–53.

Wallwork, Jo and Paul Salzman, eds. *Early Modern Englishwomen Testing Ideas*. Burlington: Ashgate, 2011.

Walmsley, Peter. *Locke's "Essay" and the Rhetoric of Science*. Lewisburg: Bucknell University Press, 2003.

Warner, William B. *Licensing Entertainment*. Berkeley: University of California Press, 1998.

Warren, George. *An Impartial Description of Surinam upon the continent of Guiana in America*. London: William Godbid, 1667.

Watt, Ian. *The Rise of the Novel: Studies in Defoe, Richardson and Fielding*. Berkeley: University of California Press, 1957.

Webb, John. *An historical essay endeavoring a probability That the Language of the Empire of China is the Primitive Language*. London: Nathaniel Brook, 1669.

Wehrs, Donald R. "*Eros*, Ethics, and Identity: Royalist Feminism and the Politics of Desire in Aphra Behn's *Love-Letters Between a Nobleman and His Sister*." *SEL* 32 (1992): 461–78.

Weil, Rachel. *Political Passions: Gender, the Family, and Political Argument in England, 1680–1714*. Manchester, UK: Manchester University Press, 1999.

Westfall, Richard S. *Never at Rest: A Biography of Isaac Newton*. Cambridge: Cambridge University Press, 1980.

Whiston, William and Humphrey Dutton. *A New Method for Discovering the Longitude Both at Sea and Land, Humbly Proposed to the Consideration of the Publick....* London: John Phillips, 1714.

Wilcox, Helen. "'Free and Easy as ones discourse'?: Genre and Self-Expression in the Poems and Letters of Early Modern Englishwomen." In Dowd and Eckerly, *Genre and Women's Life Writing*, 15–32.

Willis, Thomas. *An essay of the Pathology of the brain and nervous stock in Which Convulsive Diseases Are Treated of*. Translated by Samuel Pordage. London: J. B., 1681.

Wilputte, Earla. Introduction to *Three Novellas* by Eliza Haywood, edited by Earla Wilputte, 1–15. East Lansing: Colleagues Press, 1995.

———. "'Too Ticklish to Meddle with': The Silencing of *The Female Spectator*'s Political Correspondents." In Wright and Newman, *Fair Philosopher*, 122–40.

Wilson, Carol Shiner. Introduction to Wilson, *The Galesia Trilogy*, xv–xliii. Oxford: Oxford University Press, 1997.

———. ed. *The Galesia Trilogy and Selected Manuscript Poems of Jane Barker.* Oxford: Oxford University Press, 1997.

Worden, Brian. "The Question of Secularization." In Houston and Pincus, *A Nation Transformed*, 20–40.

Wright, Lynn Marie and Donald J. Newman, eds. *Fair Philosopher: Eliza Haywood and "The Female Spectator."* Lewisburg: Bucknell University Press, 2006.

———. Introduction to Wright and Newman, *Fair Philosopher*, 13–41.

Wright, Susan. "Private Language Made Public: The Language of Letters as Literature." *Poetics* 18 (1989): 549–78.

Yolton, John W. *A Locke Dictionary.* Cambridge, MA: Blackwell, 1993.

———. *Thinking Matter: Materialism in Eighteenth-Century Britain.* Minneapolis: University of Minnesota Press, 1983.

Young, Francis. *English Catholics and the Supernatural, 1553–1929.* Burlington: Ashgate, 2013.

Zaczek, Barbara Maria. *Censored Sentiments: Letters and Censorship in Epistolary Novels and Conduct Material.* Newark, DE: University of Delaware Press, 1997.

Zinsser, Judith P. Introduction to Zinsser, *Men, Women, and the Birthing of Modern Science*, 3–9.

———. "The Many Representations of the Marquise Du Châtelet." In Zinsser, *Men, Women, and the Birthing of Modern Science*, 48–67.

———. ed. *Men, Women, and the Birthing of Modern Science.* Dekalb, IL: Northern Illinois University Press, 2005.

Zunshine, Lisa. *Why We Read Fiction.* Columbus: The Ohio State University Press, 2006.

Index

Addison, Joseph, 108, 109, 112, 115
  The Spectator, 108, 109, 112,
    115–16, 138
Ahern, Stephen, 194n37
Alcoff, Linda, 176n27, 186n81
Allestree, Richard, 29
Alpers, Svetlana, 170
Altegoer, Diana, 186n83, 187n85,
    187n90
amatory fiction, 26, 37, 40–1, 73,
    82–3, 103, 112, 114–15, 117,
    120–2, 125–6, 144, 194n44,
    207n54
Anderson, Emily Hodgson, 38
Anderson, Misty, 77, 201n42
Areopagitica (Milton), 111
Armstrong, Nancy, 176n32
Astell, Mary, 19, 104
Atkinson, Dwight, 31, 176n28,
    184n58, 188n97, 188n104
Austen, Jane, 9, 143, 163, 170,
    215n9

Backscheider, Paula, 9, 199n27,
    203n6, 204n7
Bacon, Francis, 3, 4, 15, 16, 17,
    23, 24, 26, 30, 58, 78, 163,
    199–200n29
  Great Instauration, 3
  New Atlantis, 199–200n29
  New Organon, 24, 163,
    199–200n29
Baconian epistemology, 32, 78, 88,
    201n42
Ballaster, Ros, 5, 47, 58, 187n88,
    189n3, 194n44
Bannet, Eve Tavor, 108, 203n6,
    204n7, 205–6n23

Barker, Edward, 76
Barker, Jane, 12–13, 25, 26, 29,
    30, 71–100, 115, 127, 130,
    131, 140, 164, 167–8, 199n25,
    200n38, 201–2n48, 202n54
  "A Farewell to Poetry, with a long
    Digression on Anatomy", 77
  The Galesia Trilogy, 12–13, 71,
    73, 75, 115
  "An Invitation to my Friends at
    Cambridge", 25, 77
  *A Lining for the Patch-Work
    Screen for the Ladies*, 75, 80,
    84, 85–93, 95–7, 99
  Love Intrigues, 77–8, 79–82, 83,
    84, 94, 99, 199n27
  "On my Mother and my Lady
    W – – who both lay sick at the
    same time Under the Hands of
    Dr. Paman", 199n25
  *A Patch-Work Screen for the
    Ladies*, 77, 84, 94, 95–7, 98–9
  Poetical Recreations, 77, 200n38
Barnes, Diana, 187n92
Battigelli, Anna, 4
Baudot, Laura, 170, 183–4n50
Bayer, Gerd, 209n13
Beebee, Thomas O., 41, 193n26
Behn, Aphra, 1, 8, 9, 12, 13, 19, 29,
    35–70, 71, 73, 76, 81, 93, 94,
    127, 153, 163, 167, 168–9, 171,
    188–9n2, 189–90n4, 191n17,
    192n19, 193–4n36, 194–5n49,
    195n50, 195n54, 201–2n48,
    202n54
  "The Adventure of the Black
    Lady", 61
  A Discovery of New Worlds, 36

Behn, Aphra—*Continued*
  Emperor of the Moon, 36, 38
  The History of the Nun, 62–3, 68
  *Love-Letters Between a
    Nobleman and His Sister*, 1,
    12, 35, 36, 37, 39–61, 62,
    63–4, 66, 67, 68, 188–9n2,
    193–4n36, 195n54
  Oroonoko, 12, 35, 37, 38–9, 44,
    53, 57, 63–9, 163, 191n17
  The Roundheads, 36
  The Rover, 56
  "On Trees" (Book VI), 36
  "To the Unknown Daphnis on
    his Excellent Translation of
    Lucretius", 36, 48
  "The Unfortunate Happy Lady",
    56, 61
  The Widdow Ranter, 44, 57
Bender, John, 9, 11
Benedict, Barbara M., 37, 102,
    173–4n6, 189n3, 190n10,
    201n43
Bensaude-Vincent, Bernadette,
    191n16
Berkeley, Henrietta, 39, 57
The Black-Day (anonymous), 133,
    210n26
Blackwell, Mark, 178n45
Blondel, Christine, 191n16
Bocchicchio, Rebecca P., 204n15
body, 7, 8, 15, 27, 29, 39, 44–5,
    47–8, 51–2, 54, 56–8, 63–6,
    77, 88, 96, 103, 123–6, 152–4,
    163–4, 191n18, 192n19
  gender, 7, 18, 124
  men, 18–19
  and mind, 2, 4, 14, 15, 16–17,
    18–22, 23, 34, 47–8, 103, 125,
    128–9, 152, 163
  perception, 3, 14, 15, 17, 102–3,
    123, 164
  soul, 17, 22
  women, 29, 64
  see also self
Boone, Troy, 88

Bowden, Martha, 131, 132, 144,
    146, 211n38, 212n48,
    213n54
Bower, Anne L., 187n92
Bowers, Toni (O'Shaunessy), 73,
    131–2
Boyle, Robert, 1, 4, 7, 8, 16, 17–18,
    19, 22, 23, 27, 30–2, 35, 36,
    39, 64, 65, 66, 72–3, 76, 85,
    88, 103, 111, 188n95, 191n17,
    191–2n18
  The Aerial Noctiluca, 188n95
  The Christian Virtuoso I, 17–18,
    25, 31–2, 77
  *New Experiments and
    Observations Touching Cold*,
    191n17
  *New Experiments Physico-
    Mechanical*, 188n95
Bratach, Anne, 37, 187n87, 189n3,
    191n16, 192n19
Broad, Jacqueline, 194n46
Brown, Laura, 19
Buchholtz, Laura, 143
Burnet, Thomas, 21, 27
Butler, Joseph, 23
Butler, Judith, 10

Cajka, Karen, 203–4n6
The Canterbury Tales (Chaucer), 75
Carlile, Susan, 176–7n16, 203–4n6
Carnell, Rachel, 5, 9–10, 58,
    174n11, 184–5n60, 204–5n15,
    205–6n23
Carter, Philip, 18
Cartesian epistemology, 17, 23,
    77, 88
de Castera, Louis-Adrien Du
    Perron, 103–4, 211n31
  *La Pierre Philosophale des
    Dames*, 103–4
Catholic, Catholicism, 74, 85, 99
Cavendish, Margaret, Duchess of
    Newcastle, 4, 19, 30–1, 74, 77,
    188n95
chemistry, 156

Chibka, Robert L., 196n63
Chico, Tita, 191n16
Christianson, Gale E., 203n2
chronometer, 33, 188n105
Clark, Peter, 185n65
Clarke, Samuel, 23
Clayton, Robert, 76
Clucas, Stephen, 191–2n18
Cockburn, Catherine Trotter, 23
Code, Lorraine, 25
Collins, Anthony, 23
community, 3, 26–7, 29, 71–3, 85, 97–8, 101–2, 107, 127
Congreve, William, 9
Conway, Anne, 88
Coppola, Al, 189n3
"The Corpse Twice 'Killed'", 201–2n48
Cotes, Roger, 35
Cottegnies, Line, 189n3
Cowley, Abraham, 36
Creech, Thomas, 36, 48
Croskery, Margaret Case, 124
Culler, Jonathan, 10, 178n39, 214n6

Dalmiya, Vrinda, 176n27, 186n81
Daston, Lorraine, 163, 170, 174n8, 179n3, 179–80n12, 184n53
Davies, Tony, 193n27
Davis, Lennard, 177n34
Davys, Mary, 1, 9, 13–14, 29, 30, 126, 127–65, 167, 168–9, 171, 209n18, 211n38, 212n48, 212n49
  The Accomplish'd Rake, 1, 14, 128, 129–30, 131, 140, 141, 143, 151–63, 210n28
  The Amours of Alcippus and Lucippe, 14, 131, 140, 143–7, 149, 163, 212n48, 214n62, 214n65
  The Cousins, 131, 141–3, 148–9, 214n62
  The False Friend, 131, 141, 149, 177n35
  Familiar Letters Betwixt a Gentleman and a Lady, 130, 131–41, 151, 160, 162, 163
  The Fugitive, 129–30, 131, 141, 143–4, 149–51, 211n43, 212n49, 212n53, 214n62, 214n65
  The Lady's Tale, 130, 131, 141, 144, 147–9, 151, 214n62
  The Merry Wanderer, 129–30, 131, 141, 144, 149–51, 212n49, 212n53, 214n62
  The Northern Heiress, 152, 214n65
  The Reform'd Coquet, 128, 129–30, 131, 132, 141, 143, 151–63
  The Self-Rival, 152
  The Works of Mrs. Davys, 129, 130, 131, 144, 148, 155, 211n42
Dawson, Paul, 10–11, 169
Dear, Peter, 3, 8, 25, 72–3, 121, 173n5, 173–4n6, 176n28, 181n29, 182n35, 184n53, 185n59
De Bruyn, Frans, 128, 155, 159, 211n38, 213n54, 214n65
The Decameron (Boccaccio), 75
Defoe, Daniel, 8–9, 28
  Robinson Crusoe, 38
  A True Report of Mrs. Veal, 88
Deists, 89
Descartes, René, 4, 15, 16, 22
  Discourse on Method, 24
detachment, 1, 9–10, 18, 23–9, 33, 34, 35, 36, 52, 68–9, 82, 101–26, 127–65, 184n53
  body, 14
  isolation, 25–9, 63, 68–9, 97–8, 101–26
  mind, 15
  morality, 13–14, 23–9, 127–65
  women, 29
  see also objectivity

didacticism, 128, 144, 152, 159–61
    see also morality
Ditton, Humphry, 133
Donawerth, Jane, 187n88
Donne, John, 105, 205n20
Donovan, Josephine, 74–5,
    176–7n32, 198n19
Dowd, Michelle M., 30
Downie, J. A., 9, 177n34
Drury, Joseph, 103, 125–6, 204n15,
    206n37
Duffy, Maureen, 43, 189n3

Eckerle, Julia A., 30
Edmundson, Henry, 187n84
Eicke, Leigh A., 93
empiricism, 3, 7, 11, 77, 86, 88,
    93, 98
    see also experimental philosophy
epistemology, 2, 3–4, 9–12, 14, 15,
    17, 22–3, 25–6, 40, 72–3, 74,
    84, 85, 93, 97–8, 102, 104,
    107, 127–9, 130, 152, 163–4,
    167–71, 174n8, 179n12
    gender, 5, 6, 7, 29, 186n81
    men, 18–19
    novel, 5, 9–10, 11, 75–6, 77, 117,
    129, 168–9
    politics, 19, 35, 57–60, 132
    women, 6–7, 19
epistolary fiction, 39–48, 52–3, 95,
    60, 130, 136–40, 160, 164
epistolary form, 31–2, 187n92,
    188n95
Espinosa, Aurelio M., 201n48
Exclusion Crisis, 58
experimental philosophy, 6–7,
    23, 27–8, 31, 36, 37, 38–9,
    47, 61–2, 66, 69, 72–3, 85,
    93, 97–8, 111, 168, 185n69,
    191n17–18
    see also empiricism

Feingold, Mordechai, 23, 170,
    174n11, 181–2n32
Felton, Henry, 182–3n42

fiction, 1, 8–9, 69–70, 116–17, 120,
    129, 194n44
    see also amatory fiction;
    epistolary fiction; individual
    authors; novel
Fielding, Henry, 8, 171
Fielding, Sarah, 74
Figueroa-Dorrego, Jorge, 193n33
Finger, Stanley, 191n17, 192n19
Fitzmaurice, James, 84, 198n14,
    198–9n21, 200n32
Flamsteed, John, 101, 132–4,
    209n18
Fleming, James Dougal, 194n45
Fontenelle, Bernard de, 36, 190n4
Ford, Lord Grey, 39, 57–8
Forstrom, J. Joanna S., 182n39
Foucault, Michel, 10
Fowke, Martha, 110
Fox, Christopher, 182–3n42
framed narrative, 13, 71, 73–6, 84,
    93–8, 99, 115, 140, 141, 143,
    144, 147–51, 168–9
Frangos, Jennifer, 86, 201n43
Frankenstein (Shelley), 75
free indirect discourse, 143, 170
Funkenstein, Amos, 2, 22–3, 85,
    182–3n42

Galileo, 3, 22, 105
Galison, Peter, 163, 170, 174n8,
    179n3, 179–80n12, 184n53
Gallagher, Catherine, 176–7n32,
    197–8n11
Gay, John, 130
gender, 5–8, 18–19, 34, 56–7, 58,
    104, 110, 123–5, 126, 158,
    159–60, 162–3, 164, 168,
    180n14, 204n15
Genette, Gérard, 10, 96, 178n39
Gerrard, Christine, 28, 110
Gevirtz, Karen (Bloom), 189n3,
    212n52
ghosts, 17, 85, 87–90
    see also spirit
Gilroy, Amanda, 41

Girten, Kristen M., 203n7
Glanvill, Joseph, 72
Gleick, James, 101, 202–3n1
Glover, Susan, 132, 159, 160,
    197n11, 207n55, 210n28,
    213n54
Gold, W. L., 210n22
Golinski, Jan, 30, 174n8, 176n28,
    186n83
Goodfellow, Sarah, 189n3
Goreau, Angeline, 189n3
Gorham, Geoffrey, 173–4n6
Gowing, Laura, 19
Greene, Richard, 182–3n42
the Greshamites, 3
Gretton, Phillips, 182–3n42
Grosz, Elizabeth, 181n24
Gunn, Daniel P., 143

Hall, Rupert, 65, 203n2
Halley, Edmund, 21, 133
Hammond, Brean, 177n34
Handley, Sasha, 173–4n6
Hannay, Margaret P., 175n17
Haraway, Donna, 6, 18, 72, 186n81
Harris, Frances, 6–7
Harrison, Peter, 25, 28–9, 173–4n6,
    183–4n50, 200n35
Harrow, Sharon, 207n54
Harth, Erica, 5, 19
Hartley David, 39
Harvey, Karen, 19
Harvey, William, 26, 61, 64, 65
    *On the Motion of the Heart and
    Blood in Animals*, 26, 32, 81
Harwood, John T., 176n28
Hayden, Judy A., 6, 189n3
Haywood, Eliza, 8–9, 13, 29, 30,
    73, 93–5, 100, 102–26, 130,
    131, 138, 164, 167, 168–9, 171,
    201–2n48, 204–5n15, 205n23,
    206n37, 207n51, 211n31
    The British Recluse, 82–4, 93–5
    The Female Spectator, 28, 108–9,
    111, 114, 115
    The Invisible Spy, 103

    The Lady's Philosopher's Stone,
        103–4, 211n31
    Love in Excess, 103–4
    A Spy upon the Conjurer, 122
    The Tea-Table, 13, 14, 19, 29,
        102–26, 127, 149, 169, 207n51
    The Tea-Table, Part 2, 104,
        122–4, 207n51
Heinze, Reudiger, 95
The Heptameron (de Navarre), 75
Herman, David, 178n45
Hill, Aaron, 13, 28, 103, 104–5, 110
    The Plain Dealer, 109–10, 112
Hillarian circle, 103, 104, 110, 123
Hnath, Lucas, 202–3n1
Hobbes, Thomas, 22, 27, 36, 37,
    58, 72–3, 111, 112
Hobbesian epistemology, 88, 103,
    126, 170, 193n36
Hogan, Patrick Colm, 178n45
Honeybone, Michael, 185n65
Hooke, Robert, 1, 4, 17, 30, 35, 61,
    64, 65, 85
    Micrographia, 18, 27–8, 32, 66,
        75, 191n12, 196n60
Howard, Robert, 105
Hultquist, Aleksondra, 203n6
Hunter, J. Paul, 8, 169, 177n34,
    208n3
Hunter, Lynette, 3, 6–7, 173n5,
    174n14, 175n17
Hunter, Michael, 3, 4, 133, 173n5,
    173n6, 176n27, 176n28,
    181n32, 182n35, 200n36
Hutchinson, Keith, 200n35
Hutton, Sarah, 3, 173n5, 174n14

Iliffe, Roger, 32, 181–2n32, 203n2
The Indian Queen (Dryden), 105
individualism, 12–13, 58, 72–3, 74,
    84–5, 97–8, 105, 126, 207n54
the Invisible College, 3
irrationality, 52, 82, 104, 120,
    211n31
    men, 125, 147, 163
    women, 125

Jacobite, Jacobitism, 74, 205n23
Johnson, Esther ("Stella"), 130
Johnson, George, 202n1
Johnson, Samuel, 115
Joule, Victoria, 208n2, 210m28, 211n43

Katritzky, M. A., 191n16
Keller, Eve, 23, 30–1, 173–4n6, 174n9, 179n12
Kern, Jean B., 211n43, 213n54
King, Kathryn R., 74, 76–7, 84, 122, 198n14, 198n19, 199n22, 199n27, 201n43, 205n23, 207n51
Kramnick, Jonathan, 178n45

Lamarra, Annamaria, 193n35
Lanser, Susan Sniader, 5, 8, 10, 169, 176n30
Laqueur, Thomas, 19
layered narrative. *See* framed narrative
"learned lady", 74
Leary, John E., Jr., 15, 194n45
van Leeuwenhoek, Antonie, 33–4, 64
Lehrer, Jonah, 167
Lewis, Jayne Elizabeth, 174n11
Locke, John, 1, 5, 11, 22, 30, 127
  body, 14, 17, 21, 102–3, 124, 152, 163
  contract theory, 139
  empiricism, 11, 77, 88, 98
  epistemology, 2, 19, 22–3, 74, 92–3, 97–8, 152, 163, 164
  Essay on Human Understanding, 12, 20–2, 102–3, 128–9, 152, 167
  Newton, Isaac, 12, 20–3, 98, 181n32
  politics, 21, 58, 132, 164
  property, 132
  the self, 4, 11, 11–14, 23, 27, 85, 128–9, 152, 162, 163
  sexual contract, 139–40

Lodowyck, Francis, 187n84
Lower, Richard, 76
Lubey, Kathleen, 82–3, 114–15, 120, 203n6
Lucretius, 36, 38, 47–8, 88, 192n19

Macaulay, Catherine, 104
MacKenzie, Niall, 198n14
Maclaurin, Colin, 27
MacLean, Gerald, 31, 41
Manley, Delarivier, 8, 9, 74
Mascuch, Michael, 5, 186n82
materialism, 77, 88, 103
mathematics, 5, 21–2, 25–6, 27, 65, 73, 85, 128, 156, 184n58, 185n68, 213n58
McArthur, Tonya Moutray, 198n14
McBurney, William H., 129, 211n38, 211n43, 213n54
McCann, Edwin, 182n39, 182n42
McKeon, Michael, 8, 92–3, 177n34, 183n50, 197n11, 200n35
mechanism, 126, 170, 193n36
Medoff, Jesslyn, 76
Meeker, Natania, 177n33, 178–9n45
Merritt, Juliette, 104, 109, 112, 203n6
Messenger, Ann, 204n7
mind, 3, 15, 17–18, 72, 85, 92, 120, 126, 152–3, 163
  and body, 2, 4, 14, 15, 16–17, 18–22, 23, 34, 47–8, 103, 125, 128–9, 152, 163
Moglin, Helene, 184n60
Molesworth, Jesse, 5, 25–6, 184n60, 208n2
Monmouth's Rebellion, 39, 57–8
morality, 23–9, 34, 37, 61, 94–5, 97, 101–26, 127–65, 168–9, 178n45
More, Henry, 22
Morgan, Thomas, 28
Morris, John, 76

narrative theory, 10–11, 169
narrator, 1, 13–14, 35, 69, 71,
  127–9, 168–9
  see also body; individual works;
  omniscience
Nelson, Lynn Hankinson, 98
"The New Science", 3, 4, 12, 36–7,
  38, 69, 76, 77, 107
Newman, Donald J., 205n23
Newton, Isaac, 5, 9, 21, 85, 101,
  127, 167, 191–2n18, 192n19,
  203n2
  body, 15, 17, 39, 65
  epistemology, 85, 88, 98, 170,
    210n29
  Flamsteed, John, 132
  Locke, John, 12, 20–3, 181n32
  mathematics, 21–2, 25, 27, 65,
    73, 85, 185n68
  "Newton for the Ladies", 6
  Opticks, 12, 21, 23
  *Philosophiae Naturalis Principia
    Mathematica (Principia)*, 12,
    20, 21, 23, 25, 35, 36, 65, 66
  Royal Society, 25, 101, 132
Nicolson, Marjorie Hope, 23,
  210n29
Novak, Maximilian E., 38
novel, 2, 4–5, 6, 7–14, 15–16, 25–6,
  29, 30, 34, 58, 70, 74, 95, 99,
  102, 116, 120–2, 126, 128,
  130, 152, 159–60, 164–5,
  167–71
  see also amatory fiction;
  didacticism; epistolary fiction;
  framed narrative; realism

objectivity, 7, 52, 65, 67–8, 143,
  163, 179n12, 184n53
O'Donnell, Mary Ann, 189n4,
  190n5, 191n17, 195n54,
  210n21
Oldenburg, Henry, 33, 111
omniscience, literary, 2, 10–11,
  94–5, 99, 102, 127, 131, 164–5
omniscient narration, 143, 169

omniscient point of view, 10–11,
  13–14, 126
Orr, Leah, 122, 188n2, 204n11
the Oxford group, 30, 76

paralepsis, 94–5
Parker, J., 210n26
Patchias, Anna C., 124
Pateman, Carol, 139
Pears, Iain, 71
Pearson, Jacqueline, 69
Pechey, John, 213n58
Perception, 3, 11, 14, 15, 17, 20–2,
  24, 25, 26, 29, 51, 54, 56, 57,
  72–4, 84, 88, 98, 112, 153
  see also senses
Perronet, Vincent, 23
Perry, Ruth, 193n26
Pettit, Alexander, 108, 111, 124,
  207n50, 207n51, 207n53
Petty, William, 21–2
philosophical revolution, 1–4, 11,
  14, 15–34, 35–9, 58, 71, 75,
  76, 100, 101–2, 105, 136, 164,
  167–71
  women, 5–7, 102, 168–9
Philosophical Transactions, 21–2,
  31–3, 61, 184n58, 188n95,
  188n97
Pitt, Robert, 213n58
Platonists, 3, 89
politics, 9–10, 18–19, 21, 56–8,
  106, 131–2, 134–6, 140,
  167–8, 188n2, 193n39,
  194n44, 212n48
  epistemology, 18–19, 35, 40
  individualism, 74, 126
  natural philosophy, 3, 22, 30,
    91, 93
  self, 12, 34, 40, 45–6, 56
  see also Jacobite; Monmouth's
    Rebellion; Whig
Poovey, Mary, 177n34
Pope, Alexander, 130
Porter, Roy, 173n1, 179n5, 179n12,
  182n39, 182n42

Potter, Elizabeth, 6, 18, 19, 31, 98, 173–4n6, 186n81
Potter, Tiffany, 103, 204n15
Power, Henry, 37–8, 191n12
Prescott, Sarah, 129, 176–7n32, 208n7
Pritchard, Will, 186n80
Protestant, Protestantism, 85, 93

R. F., 187n84
rationality, 4, 25, 35, 36, 88, 92–3, 98, 120, 157, 158
men, 7, 19
women, 19, 164, 167
realism (literary), 25–6, 58, 74, 128, 184n60, 208n2
reason, 11, 14, 22, 23, 82, 92, 104, 119–20, 147, 155, 157–8
Reed, Walter L., 197n11
reformed language, 30
Regan, Shaun, 177n34
Reiss, Timothy J., 2, 18, 24–5, 28–9
Revolution of 1688, 58, 74, 93
Richardson, Alan, 178n45
Richardson, Samuel, 8, 171
Richetti, John, 199n27
Riley, Lindy, 140, 210n28, 211n37, 214n65
Rivero, Albert J., 58, 194n48
Rose-Wiles, Lisa, 210n22
Rousseau, George S., 5
Royal College of Physicians, 213n58
Royal Society, 3, 15, 21–2, 30, 31–2, 37, 40, 72, 78, 96, 106–7, 111, 113, 135
Fellows, 21, 25, 27, 36, 58
Newton, Isaac, 25, 101, 132
women, 6, 19
Royle, Nicholas, 10
Rubik, Margarete, 44, 57, 63, 68
Rye House Plot, 188n105

Salmon, William, 213n58
Salzman, Paul, 186n82
Saxton, Kirsten T., 205n23
Schiebinger, Londa, 5, 175n17

Schneider, Gary, 193n26
"scientific" revolution. *See* philosophical revolution
Scott, Joan Wallach, 19
Scott, Walter, 9
senses, 17, 20, 69, 119
see also perception
Shadwell, Thomas, 36, 135
The Virtuoso, 134
Shaffer, Simon, 3, 18, 26–7, 30–1, 66, 72–3, 78, 111, 173n5, 173n6
Shapin, Steven, 3, 18–19, 26–7, 30–1, 66, 72–3, 78, 111, 125, 173n5, 173n6, 180n14, 182n39, 184n56, 185n68
Shapiro, Barbara, 183n43, 183–4n50
Shaw, Narelle, 215n9
Sheridan, Patricia, 182n39
Shevelow, Kathryn, 180n20
Slare, Frederick, 27
Smith, Amy Elizabeth, 31, 40
Smith, George, 213n58
Smith, Justin E. H., 200n35
Snider, Alvin, 47–8, 189n3, 189–90n4, 190n6, 192n19
sociability, 27, 109, 114–16
soul, 2, 16–17, 22–3, 182n42
Spedding, Patrick, 207n51
Spencer, Jane, 131, 132, 134, 162, 176n32, 198n19, 211n32
spinster, 74
spirits, 3, 85, 87–9, 201n39
see also supernatural
Spolsky, Ellen, 10
Sprat, Thomas, 36, 43
St. Germain, 84, 201n43
Staves, Susan, 162, 208n2, 213n54
Steele, Richard, 108, 109, 112, 115
Stefanson, Donald Hal, 211n38, 211n42, 211n43, 213n54
Stephens, Joanna, 39
Stillingfleet, Edward, 27
Strawson, Galen, 180n20
Stubbe, Henry, 190n11

subjectivity, 4–5, 21, 23, 41, 95, 170
supernatural, 17, 82, 85, 86–93, 98,
    201n39
Swenson, Rivka, 98–9, 198n14,
    198n19
Swift, Jonathan, 130

Taylor, Charles, 2, 184n51
tea table, 104–11, 116, 121–2, 124,
    138, 205n23
Thompson, Helen, 177n33,
    178–9n45, 189n3, 193n36,
    206n37
Tierney-Hynes, Rebecca, 8
Todd, Janet, 40, 44, 176–7n32,
    189n3, 189–90n4, 195n54,
    197n11, 208n5
Turner, Martha A., 177n33

Verhoeven, W. M., 41
Vermeule, Blakey, 178n45

Wahrman, Dror, 2, 18, 170, 179n3,
    179n12
Wakely, Alice, 131–2, 134, 214n65
Wallwork, Jo, 186n82, 187n88
Walker, Margaret, 146, 212n48
Walmsley, Peter, 22, 181n32,
    182n39, 186n83, 194n43
Warner, William B., 177n34,
    197n11
Warren, George, 37, 191n17
Watt, Ian, 8, 74, 197n11

Webb, John, 187n84
Wehrs, Donald R., 42, 43
Weisl, Angela, 201n48
Westfall, Richard S., 181n32,
    202–3n1, 203n2, 209n18
Whig, 28, 57, 131–2, 134–40, 164,
    205n23
Whiston, William, 133
widow, 62, 86, 91, 130, 149,
    212n52
Wilcox, Helen, 40–1
Willis, Thomas, 76
Wilputte, Earla, 203n6, 205n23
Wilson, Carol Shiner, 198n19,
    199n27, 202n54
witness, 3, 26, 31, 47, 54, 61, 63–4,
    67, 78, 96, 98, 107, 112, 117
  modest witness, 26, 36, 64, 72,
    75, 89, 122
  virtual witness, 8, 66, 78, 89
Wollstonecraft, Mary, 10, 104
Worden, Brian, 183n50
Wren, Christopher, 76
Wright, Lynn Marie, 205n23
Wright, Susan, 193n26

Yolton, John W., 182n39
Young, Francis, 173–4n6, 201n39,
    201n43

Zaczek, Barbara Maria, 193n26
Zinsser, Judith P., 3, 20, 173n5
Zunshine, Lisa, 178n45

Printed in the United States
By Bookmasters